Experiences in Liberal Arts and Science Education from America, Europe, and Asia

William C. Kirby • Marijk C. van der Wende
Editors

Experiences in Liberal Arts and Science Education from America, Europe, and Asia

A Dialogue across Continents

palgrave
macmillan

Editors
William C. Kirby
Harvard University
Cambridge, Massachusetts, USA

Marijk C. van der Wende
Amsterdam University College
Amsterdam, The Netherlands

Assistant editor
Austin X. Volz

ISBN 978-1-349-94891-8   ISBN 978-1-349-94892-5   (eBook)
DOI 10.1057/978-1-349-94892-5

Library of Congress Control Number: 2016947371

Printed on acid-free paper

This Palgrave Macmillan imprint is published by Springer Nature
The registered company is Nature America Inc. New York

PREFACE

A renewed and global dialogue on liberal arts and sciences education is engaging many regions and countries, who seek to gain from this model in response to the twenty-first-century requirements for excellence and relevance in undergraduate education. This is illustrated by inspiring examples of experimentation, reform, and international cooperation with liberal arts and sciences models across continents.

This book builds on a small-scale seminar co-sponsored by the Harvard China Fund and Amsterdam University College, hosted by Harvard Center Shanghai on May 20–21, 2015. It brought leaders and scholars in the field of liberal arts and sciences education from around the world together to discuss regional trends and models. They focused on how this model can be implemented in different contexts and across academic cultures, structures, and traditions. They asked how the model relates to the changing experience of teaching and learning and to the contextual role of cultures and values. Various international innovations, start-ups, and major international collaborations between American, European, and Asian institutions were explored so as to understand the opportunities and the challenges for each context in developing liberal arts and sciences education. Trends were discussed with a view to system-level impact, secondary education structures, and demands from the labour market. A specific focus was placed on developments in, and cooperation with, China, which could gain from this model in terms of global integration and influence, while sustainable success would require substantial governance reform.

A selection of these contributions is collected here and aims to provide a basis for a continued cross-continental dialogue in the years to come. Joint aspirations and mutual inspiration support the essential goals of liberal arts and sciences education today—educating the whole person for a global world.

We would like to express our gratitude to the authors for sharing their great vision and experiences, demonstrating how the world's most outstanding institutions are leading the way in liberal arts and sciences education. We thank Austin Volz for his excellent assistance in editing this volume, and we also thank Julia Cai and the staff of the Harvard Center Shanghai for their professional support during the seminar.

William Kirby

Marijk C. van der Wende

CONTENTS

Notes on Contributors

Li Cao is Professor of English and American Literature and deputy Director of Liberal Education at Tsinghua University, China. She also serves as vice president of the Chinese Association for the Studies of Literature in English and vice president of the International Federation for Modern Languages and Literatures (FILLM).

Dirk Van Damme is head of division in the Directorate for Education and Skills at the OECD in Paris, where he leads the Centre for Educational Research and Innovation and the Indicators of Educational Systems Programme. He previously served in education policy in the Flemish part of Belgium and as an expert for various international organisations.

Nicholas B. Dirks is the Chancellor of the University of California, Berkeley. An internationally renowned historian and anthropologist, Dirks is also a committed advocate for accessible, high-quality public higher education in the liberal arts and sciences.

Nicholas Eschenbruch was the founding Director of Education of University College Freiburg from 2011 to 2015. By training, he is a social anthropologist and historian with a research focus on twentieth-century medicine.

Hans-Joachim Gehrke Professor emeritus, is representative of the Rector for the University College Freiburg. From 1982 to 2008, he was Professor of Ancient History at the University of Würzburg, Free University of Berlin, and University of Freiburg, and from 2008 to 2011, president of the German Archaeological Institute.

Carl Gombrich is the programme Director of University College London's Arts and Sciences BASc degree. Carl has degrees in mathematics, physics, and philosophy and was formerly an opera singer. He is an international speaker on liberal and interdisciplinary education, and a member of the British Academy Working Group on Interdisciplinarity.

William C. Kirby is T.M. Chang Professor of China Studies at Harvard University, Spangler Family Professor of Business Administration at the Harvard Business School, and Harvard University Distinguished Service Professor. A historian of modern China, Professor Kirby's work examines China's business, economic, and political development in an international context.

Pericles Lewis is the Founding President of Yale-NUS College, which has been cited as a model for reinventing residential liberal arts and sciences education in the context of twenty-first-century Asia. He has served as an advocate for liberal education in Singapore, in the USA, and internationally.

Bryan Penprase is Professor of Science at Yale-NUS College, where he serves as the inaugural Director of the Yale-NUS Teaching and Learning Center, and also the Frank P. Brackett Professor of Astronomy at Pomona College. At Yale-NUS, he is developing interdisciplinary science courses, and working on global liberal arts initiatives.

Gerard Postiglione is Chair Professor of Higher Education and Associate Dean for Research in the Faculty of Education, The University of Hong Kong.

Paul Sterzel studied political science, economics, and history in Germany and France. A research career phase focusing on globalisation of science and technology policy led him towards more practical aspects of science and education management. Since 2011, he is managing director at University College Freiburg.

Marijk C. van der Wende was the founding Dean of Amsterdam University College (2007–2014) and visiting scholar at Harvard University's Fairbank Center for Chinese Studies (2015). She was since appointed as Dean of Graduate Studies and Professor of Higher Education at Utrecht University.

CHAPTER 1

A Global Dialogue on Liberal Arts and Sciences: Re-engagement, Re-imagination, and Experimentation

William C. Kirby and Marijk C. van der Wende

Abstract This chapter provides both an introduction and overview of the book. It discusses the background and history and conceptual arguments for liberal arts and sciences education for the global twenty-first century, as well the various trends of re-engagement with the liberal arts model in the USA (notably Harvard University) at the start of the century, the re-imagination of the model in Europe (more specifically by the Amsterdam University College experience) and the various forms of experimentation in China as well as the conditions under which these could best succeed.

Keywords liberal arts and sciences education • 21st century • Under-graduate education • global dialogue • China • Europe

In a world of higher education increasingly consumed either with the growth of professional education, particularly in business, or with "rankings," or "league tables" of research universities and the quest for "world-

W.C. Kirby (✉)
Harvard University, Cambridge, MA, USA

M.C. van der Wende
Amsterdam University College, Amsterdam, Netherlands

1
W.C. Kirby, M.C. van der Wende (eds.), *Experiences in Liberal Arts and Science Education from America, Europe, and Asia,*
DOI 10.1057/978-1-349-94892-5_1

class" institutions, we seek in this volume to talk about undergraduate education of both a traditional and very modern sort: about education, teaching, and the surprisingly enduring and now expanding conceptions of the liberal arts and sciences in a twenty-first-century education.

The global debate on the liberal arts and sciences has moved from its European origin and out of the North American academic landscape to engage many regions and countries, including China, which seeks to gain from this model in terms of global integration and influence. This is illustrated by inspiring examples of experimentation, reform, and international cooperation with liberal arts and sciences models.

This book highlights the visions and experiences of international leaders in the field of liberal arts and sciences education from around the world. The authors discuss regional trends and models, several with a specific focus on why this model seems to respond to twenty-first-century requirements for excellence and "real world" relevance in undergraduate education. Taken together, the essays explore how liberal arts and sciences curricula can be implemented in different national contexts and across a broad range of academic cultures, structures, and traditions. They investigate how teaching and learning experiences may vary in the context of different cultures and values. A variety of international innovations, startups, and major international collaborations between American, European, and Asian institutions are explored in order to understand the opportunities and the challenges for China in developing liberal arts and sciences education. The authors have reviewed and evaluated trends with the aim of making impact across whole systems of higher education, with implications also for secondary education before university and the demands of labor markets after graduation.

Let us start with some reflections on what education in the liberal arts and sciences entails. The debate on these issues goes back minimally to those of the nineteenth century between proponents of the Humboldtian ideal of *Bildung* (the education of the whole person) as distinct from *Übung* (more practical training), differences that are phrased differently across the world—in China, for example, as the distinction between a broad conception of education (*jiaoyu* 教育) and a narrower, repetitive one of training (*xunlian* 训练).

Wilhelm von Humboldt, who founded the University of Berlin in 1810, envisioned an education that was broad and deep, rooted in fields in which he was deeply learned—history, classical literature and languages, and linguistics—giving citizens the capacity for self-cultivation and individual development in society. He (as notably did also his brother, Alexander) believed also in scientific research, the creation of new knowledge, and he aimed to create an institution in which teaching and research would be integrally connected, with teaching rooted in research, in an institution free from preordained orthodoxies.

The University of Berlin is the ancestor of all modern and contemporary research universities. Humboldt's ideals have been reflected in the principles of *Lehrfreiheit* and *Lernfreiheit* (freedom to teach and freedom to learn) that have been at the heart of modern conceptions of academic freedom.

Yet the modern research university today is much larger, much more professionalized, much more focused on research, and much less focused on teaching than any institution that Humboldt could have imagined. In Germany, by the early twentieth century, a process had begun to separate out high-level research institutes from universities (named initially for Kaiser Wilhelm II, now for the physicist Max von Planck). In the USA, which in the twentieth century became home to many of the world's leading research universities, the growth of stand-alone or in-house research institutes with little or no formal teaching has been a growing feature of higher education. To give one example that we know well: Harvard University, for example, grew from a provincial college to first a national then international research university, faculty time and energy was inevitably drawn toward graduate education and professional engagement far removed from the undergraduate classroom.

The challenges faced by Harvard and by many of the universities described in this volume have commonalities, but central to them is an overwhelming concern with committing, or re-committing, these institutions and their faculty to the challenges of a broad, *undergraduate* education in a world seemingly dominated (as Humboldt's was not) by pure research and by onrushing advances in science and technology.

We see in many of the examples discussed in this volume an international commitment to both general and liberal education in the broadest sense. What is "liberal education?" As Montaigne wrote: "Among the liberal arts, let us begin with the art that liberates us." Montaigne was referring to a process whereby previously unexplored beliefs and values are challenged as well as unsuspected dimensions of the self, discovered and nurtured in order that students may become "wiser and better" for themselves and for society. Liberal education presumes that a broad education will liberate the individual by offering opportunities for foundational knowledge, reflection and analysis, artistic creativity, and an appreciation for the precision of scientific concepts and experiments.

The American tradition of liberal arts and sciences education has been most robust in a (until recently) unique institution: the independent college of liberal arts. Institutions such as Oberlin, Williams, Carleton, Reed, and many others have always employed first-class scholars, normally trained at leading research universities, but their institutional focus has been exclusively on undergraduates. Theirs has been the strongest commitment to the idea of liberal education: educating the whole person, and not just training the specialist. They resist pressures for early specialization and professionalization. Professional education may be the proud tradition of many great universities, but it has not been the fundamental mission of the American liberal arts college. While their students will have devoted some significant part of their time to special and concentrated learning, they aim to graduate having developed their intellectual, artistic, moral, civic, and scientific capacities as independent thinkers with a lifetime of learning still before them.

The challenge of leading American research universities is different from that of independent liberal arts colleges. Those that have grown from the foundations of famous undergraduate colleges (e.g. Harvard, Yale) to become large and complex research institutions have the difficult task of trying to keep the undergraduate enterprise at the center of a big university. How do these research-driven universities put the energy of leading scholars back on undergraduates? How do highly selective institutions prepare their students to enter a globalizing world of national conflicts, of scientific advance, of political choice and economic uncertainty, of artistic imagination and cultural repression? There is, of course, no "one-size-fits-all" educational menu for such alternative futures, but the last decade has been one of debate and renewal of the liberal arts in many such leading American universities.

In the spring of 2007, the Harvard faculty approved a new General Education curriculum for Harvard College, after several years of drafting and seemingly endless discussion. When it passed with near unanimity, the faculty was told about the famous 1924 debate in the Chinese Communist Party about joining the Nationalists in the first United Front. The minutes of that meeting were recorded thus: "The resolution passed unanimously, even though many comrades were opposed." Revising entrenched systems of undergraduate education is never easy, and there is no perfect model. The Harvard effort puts greater stress on internationalization, scientific and technological literacy, and new communities of learning in smaller settings than had its predecessor efforts. Above all, it tried (and so far has succeeded) in having a new generation of faculty re-engage in undergraduate education and to create courses and departmental curricula for which they had intellectual ownership and responsibility.

The American reinvestment in the liberal arts and sciences is perhaps not surprising given the long history of institutions based on this concept, even if it is by no means the centerpiece of the large majority of US universities. Much more surprising, in our view, is the resurgence of liberal arts and sciences elsewhere: in Europe and in China in particular.

In recent decades, European universities have adapted some of the formal structures of perceived American models, such as the US baccalaureate. Distinct undergraduate (bachelor) and graduate (master and doctorate) degree cycles were introduced following the Sorbonne (1998) and Bologna (1999) declarations. By reinstituting the bachelor as an educational phase in its own right, these structural reforms facilitated the (re-)emergence of liberal arts and sciences programs in Europe (Van der Wende 2011).

Many of the ideals of what has since become known as the Bologna Process carry with them the promise of making higher education in Europe a continental-wide enterprise, thereby facilitating student, faculty and staff mobility, which has been growing since the introduction of the Erasmus Program in 1987. Such mobility is important in competing, and in cooperating, with continental-sized systems of higher education, such as in the USA and China.

But while there is some emulation of the current American concept of the baccalaureate, until recently, European universities have appeared less interested in the educational values that have defined the Bachelor of Arts degree in many American colleges, which stress a broad undergraduate education in the liberal arts and sciences. If one looks at the documents of the Bologna, Prague, Berlin, Bergen, and other meetings, there is enormous attention paid to research, to funding, and to math, science, and technology, and precious little to teaching, to citizenship, and to valuing the broad and deep education of the next generation of Europe's citizens. The "key competencies" for lifelong learning recommended by the European Parliament in 2006 quite appropriately include language learning; information and communication technologies; and math, science, and technology. But nearly absent are the humanities, the multidisciplinary study of other cultures and religions, and education in moral reasoning and philosophy. Even the "harder" social sciences seem short-changed.

It may be that any Europe-wide reform must be limited, given the restricted mandates of European institutions[1] and the sovereignty of European member states in the domain of (higher) education. But it cannot be denied that such EU policies do lean heavily on economic rationales. An underlying human capital approach expects higher education to propel economic growth and spur competitiveness in the global knowledge economy, leading to a utilitarian focus on skills, rather than on values that would underpin European identity and citizenship as a basis for further social and political integration.

Now, however, with the emergence of truly innovative "university colleges," European universities, led by those in the Netherlands, seek to do the Americans one better by bringing together the separate strengths of the stand-alone liberal arts college and of the large research university.

Driven by the need to overcome the disadvantages of early and overspecialization, to differentiate the massified and overly egalitarian European higher education systems, and to meet employers' demands for well-rounded graduates, various leading universities went back to their roots. There they recovered the origin of the European university which taught the *artes liberales*, including the *trivium* (literary arts; grammar, logic, rhetoric) and the *quadrivium* (mathematical arts; arithmetic, geometry, music, astronomy). Re-imagined, twenty-first-century-oriented versions of this model now re-enter undergraduate education at the center of large and complex research universities and help to re-balance their teaching and research missions by sheltering undergraduate teaching from the pressures of research performance, rankings, and reputation race. These initiatives draw the energy of leading scholars back to undergraduate teaching, re-committing them to the challenges of a broad education that prepare students for a globalizing and evermore complex world.

Amsterdam University College (AUC) is a prominent example of the Dutch model, which is typically a highly selective honors college that teaches a three-year liberal arts and sciences bachelor and has a radically international ethos and community. The model combines the virtues of a small-scale residential college with the resources and facilities of a large research university. Ten such university colleges have been established, following the first such initiative launched in 1998 by Utrecht University (University Colleges Deans Network [UCDN] 2014). They are all fully owned by a Dutch research university and are granted a privileged status in the higher education legislation allowing for additional funding and more autonomy. The demand for this model is confirmed by the exceptionally strong growth in both domestic and international applications in comparison with other undergraduate programs.

AUC was established in 2009 as an excellence initiative jointly undertaken by the University of Amsterdam and VU University Amsterdam.

They joined forces to create a liberal arts and sciences program, based on the vision that the leaders of the future will have to work together across the boundaries of nationalities, cultures, and disciplines, in order to be successful in the globally engaged and culturally diverse society of the twenty-first century.

AUC's mission, "Excellence and Diversity in a Global City," reflects the belief that both excellence and diversity matter, as both competition and cooperation are key to success in a globalized world. Leadership does not only require excellence but also the understanding and valuing of diversity (Van der Wende 2013a). AUC students "are offered what US liberal arts colleges can only envy: access to laboratory-based research projects and the lab facilities of a research university," as emphasized by the president of the European Research Council, who also stressed: "AUC seeks to link the parts of our *globus intellectualis* that seem to have become separated, much like oceans dividing the continents. Reconnecting the natural sciences—physics, chemistry, and the life sciences—with the humanities and social sciences" (Nowotny 2012).

The AUC curriculum was designed from scratch based on orientations and considerations quite similar to that of Yale-NUS College as described by Lewis in Chap. 4 and Penprase in Chap. 5. AUC allows students to focus on "big questions in science and society" from a multi and interdisciplinary perspective. Because: "Snow was right.[2] A complete education should be a multidimensional experience, since students, teachers, schools, and research are all multidimensional" (Dijkgraaf 2009, p. 22). AUC's curriculum reconnects the sciences, humanities, and social sciences by means of a rigorous academic core that ensures the development of strong analytical and quantitative skills as well as foreign language and intercultural competence in students from all majors. It offers all students ample opportunities to focus on science and science-related subjects, because: "There are many great crises or challenges facing the world: food, energy, climate, pandemics, all driven by globalization. And many of our students will later be in a position to make important decisions, whether in business, government, policy or academia. The scientific way of thinking and approaching life could be valuable if not crucial for their success" (Dijkgraaf 2009, pp. 23–24).

The university college model allows research universities to overcome the fragmentation of disciplinary silos and to create a more flexible, open, and intellectually challenging space for undergraduate learning, simultaneously allowing for student choice and focusing on interdisciplinary and international themes. The model attracts a larger number of (also female) students to science subjects and has gained strong support from employers, meeting their demands for well-rounded graduates or "T-shaped professionals" with quantitative and qualitative abilities, soft skills, and STEM competences.

In Chap. 9, Van Damme analyzes this approach of "transcending discipline boundaries" to liberal arts and sciences education to show that it holds the promise to provide better answers to the needs and demands of the twenty-first century. However uncertain labor markets and skills demands may be, with loosening ties between degrees, skills, and jobs, in Chap. 6, Gombrich argues for the re-emergence of polymathy and generalism as educational ambitions related to the future of work.

The university college model inspired and coincided with initiatives in other European countries, including Germany, presented by Eschenbruch, Gehrke, and Sterzel from University College Freiburg in Chap. 7, and England at UCL, an illustration of Gombrich's argument for polymathy in Chap. 6. Yet the model is no panacea. While more internationally oriented and better facilitated than most US stand-alone liberal arts colleges, it remains a small-scale, cost-intensive solution. To make such a liberal arts model mainstream throughout the entire undergraduate phase, as Dirks describes of major US research universities in Chap. 8, represents an immense challenge for the overly regulated public European universities. These universities may also see a limited scope for general education in tertiary education as it has such a strong foothold, and usually lasts longer, in secondary education.

Thus, the model is so far an elite option. The UK examples all belong to the Russell Group of universities. Freiburg was among the first to benefit from the very competitive German Federal Excellence Initiative, and the Dutch colleges are all owned by the country's highly ranked research universities. As such, these initiatives contribute to the differentiation of higher education systems with the advantage of breeding excellence, seen as essential in the face of global competition.

But it also confirms Lewis's point regarding eliteness and access as its pitfall (Chap. 4). Globalization generates both new generations of cosmopolitans, who increasingly opt for international and bilingual education, as well as immigrants who face challenges in bridging the cultures and languages of old and new home countries. Colleges aiming to prepare students for a global future should include both these populations.

But even those that value diversity as a broad and inclusive concept, who feel that a white middle-class student population would be inadequate and inappropriate for their mission, and are aware of the role that cultural and social capital may play in admission processes and in residential obligations, may find it easier to attract international students than local minorities (AUC 2012). The latter group may have a specific preference for professional degrees for reasons of social mobility, but middle-class students likewise seek these professional options after their undergraduate degree. Establishing good connections to (also heavily regulated) professional graduate programs is therefore another challenge in Europe, with hitherto more success in medical and engineering schools than in law and teaching. Fortunately, employers are able to mobilize support for both aspects as they value diversity and broader skills profiles in their workforce.

Experiments in liberal arts and sciences education in Europe are building on a strong (if at times dimly recalled) foundation in humanistic traditions. There are parallels in Asia, as we see in this volume: in India and notably in Singapore (see the essays by Penprase in Chap. 5 and Lewis in Chap. 4); in Hong Kong, where every one of the eight government-supported universities has created a new form of general education (see the essay by Postiligone in Chap. 2); and also recently in China (the essay by Cao Li on experiments in liberal arts and sciences education at Tsinghua University, once known as "China's MIT." See Chap. 3)

Let us focus here on the Chinese scene. China is home to arguably the world's oldest, continuous tradition in the humanities, and a commitment to an education rooted in values—a concept recalled in Cao Li's essay by the current phrase, quality education (*suzhi jiaoyu* 素质教育), to define new forms of general education.

Until the twentieth century, the study of Chinese tradition, defined by officially sanctioned texts, not only defined what it meant to be educated but also served as a path to officialdom, and to wealth and influence. The famous imperial examination system, which lasted in one form or another for nearly a millennium before 1905, brought the empire's most learned men—only men—into the service of the state—not because they had been trained in statecraft or tax collection but because they had deeply studied what we would today call the "humanities": because they had studied, memorized, chanted, and metaphorically consumed the classics, and they would, in office, act according to the principles of human behavior that the *Analects, Mencius*, and other great works set out.

Perhaps never has there been a higher academic ideal: good people embarking on the living study of great books in order to do good work in society.

This was the ideal, of course never fully realized in practice, and the ordeal of studying to be a scholar-official was a tortuous one, captured satirically in Wu Jingzi's eighteenth-century novel, *The Scholars* (*Rulin waishi* 儒林外史, 1992) and I. Miyazaki's scholarly classic, *China's Examination Hell* (1976). There were limits to this system, which became painfully obvious in China's nineteenth-century encounter with the West: the absence of the study of mathematics, of science, of practical affairs, did not mean that the Empire was thereby better governed or better defended.

When the ancient examination system ended overnight, in 1905, it would be replaced by institutions shaped explicitly on international models, from technical institutes to liberal arts colleges to comprehensive universities. Every major Chinese institution today has an intellectual and indeed architectural foundation that is international in origin. The first full campus of Tsinghua University, for example, may be mistaken for an American Midwestern university, for it was the president of the University of Illinois at Urbana-Champaign who convinced the American president, Theodore Roosevelt, to remit Boxer Indemnity Funds to support the founding of Tsinghua. In the 1950s, during an era of Stalinist influence, Tsinghua would be reconceived, in architecture and curriculum, on the model of Moscow State. Peking University sits today on the original, American-designed (albeit in Chinese style) campus of Yenching University, a great private college of the pre-Communist era. Nanjing University has two international parents: Jinling College, a women's college that partnered with Smith College in the USA; and National Central University, founded by the Chinese National Government in 1930 and modeled on what was then the most prestigious university in the world, the University of Berlin. (To emphasize this point, a model of the Brandenburg Gate serves as an entryway both to the original grounds of National Central University and to Nanjing University's new campus.) Particularly in the Republican era (1912–1949), China built one of the world's most dynamic (if small) systems of higher education comprised of institutions Chinese and foreign, public and private. These would be Sovietized in the 1950s and nearly destroyed by Mao Zedong's "cultural revolution" in the 1960s, but the memory of excellence and internationalism has helped to fuel the more recent regeneration and growth of Chinese higher education.

What did this history mean for the liberal arts and sciences? Well before the Communist takeover, Chinese education at all levels began to drift strongly toward the study of those subjects that could defend China in a hostile world, and bring about a return to "wealth and power" (*fu qiang* 富强)—primarily through mathematics, science, and engineering. Within a decade of the end of the old imperial examinations, the moral foundation of both Chinese government and culture, Confucianism, would come under a withering attack during the "May Fourth Movement" of the late 1910s and 1920s, even as China's leading intellectuals of that era were deeply educated in both the Chinese classics and modern international disciplines. After 1932, American models of general education were gradually replaced by European-style, discipline-specific training. By 1949, when the mainland fell to the Communists, less than 10% of graduates of Chinese public universities graduated with degrees in humanistic disciplines. The Communists then took that number to the vanishing point.

As study of the humanities declined, education became at once more practical and more political. The dream of Chinese leaders from Sun Yat-sen, the first provisional president of the Republic of China, was to physically engineer and indeed a new citizenry. This was the dream a government of technocratic expertise, capable of "reconstructing" (*jianshe* 建设) China with roads, railroads, and dams—a government of huge engineering ambition, as seen in the Three Gorges Dam project, first conceived by Sun Yat-sen in the 1920s, and now built by the governments of Jiang Zemin and Hu Jintao. In recent decades, nearly every recent member of the Standing Committee of the Politburo of the Chinese Communist Party—the seven to nine or more men who run the country—has had training in engineering. China today is home to the largest pool of engineering talent in the world.

The two Leninist party-states that have ruled China from 1927 to the present have put a strong political mark on higher education. The National Government of the Guomindang (Nationalist Party) aimed at once to nationalize or otherwise regulate higher education and to "partify" university curricula (*see* Yeh Wen-hsin, *The Alienated Academy* 1990). At the same time, "culture" and the arts were to be subordinated to the purposes of the developmental state. First, under Chiang Kai-shek's New Life Movement in the 1930s and, devastatingly, under Mao Zedong's Cultural Revolution of the 1960s; art, culture, and the humanities were mobilized for the purposes of the party-state. As Mao Zedong put it in the 1940s, even before he seized power, literature and art were to be defined as "the artistic crystallization of the political aspirations of the Communist party." And these traditions continue in part today. President Xi Jinping has cited the need for enhanced Communist Party control of universities, and he has recently echoed Mao's call for literature and the arts to follow the lead of the Chinese Communist Party: "Contemporary arts must also take patriotism as a theme, leading the people to establish and maintain correct views of history, nationality, statehood, and culture while and firmly building up the integrity and confidence of the Chinese people" (Canaves 2015).

At the same time, there are countervailing trends and experiments. "General education" (*tongshi jiaoyu* 通识教育) is now the cornerstone of curricular reform in leading universities throughout the People's Republic, as well as in Hong Kong and Taiwan.

In the People's Republic of China, universities have long had general education programs of a certain sort: required classes (*bixiu ke* 必修课) in Marxism–Leninism–Mao Zedong Thought, at least one version of which is now available globally as an online course (*see* Hernandez 2015). Like required courses everywhere, students loathe and endure these *bixiu ke*. Over the past 15 years, however, mainland universities, together with those in Hong Kong and Taiwan, have competed to introduce general and liberal education programs that open opportunities for learning across the humanities and social sciences.

The expansion of general education in Chinese university curricula has taken place in new institutions (e.g. Fudan College as the liberal arts college to which all Fudan University undergraduates belong) or it may be embedded in distribution requirements. Either way, it is a sign that pace-setting Chinese universities—the ones that disproportionately educate future political leaders—now assert that China's next generation of leaders should be broadly educated in the humanities and social sciences as well as in the sciences. In 2001, Peking University inaugurated the Yuanpei Program (now Yuanpei College), named for Peking University's famous German-educated chancellor of the early twentieth century, the philosopher Cai Yuanpei, as part of a broad reform of undergraduate education to foster "a new generation of talented individuals with higher creativity as well as international competence so as to meet the needs of our present age." Tsinghua University's School of Economics and Management, under the leadership of Dean Qian Yingyi, who received his doctorate at Harvard and holds a professorship at Berkeley, has implemented among the most imaginative program in liberal arts and general education to be found in any Chinese university—and this in a professional school. Cao

Li's article in Chap. 3 of this volume recounts Tsinghua's experiments with moral or "quality" education (*suzhi jiaoyu* 素质教育) and now the founding of Xinya College, a residential college devoted to general education. Renmin University in Beijing, founded as the "People's University" on a Soviet model, now houses several of China's leading centers of classical studies and Chinese history. To these must be added the return of international institutions (now as joint ventures): the liberal arts college opened by New York University in the form of NYU-Shanghai, and plans for a liberal arts college in the 200-acre residential campus of Duke Kunshan University, outside of Shanghai.

Perhaps Chinese educational leaders, at least in the elite institutions, believe that they need to do this, in part because, in China, as in the USA and Europe, all the pressures are in the opposite direction: on the part of students, who too single-mindedly pursue their careers, and, on the part of faculty, whose careers and interests are ever more specialized, and for whom good teaching is seldom rewarded—leading to a situation in which students and faculty interact on ever narrower ground. Perhaps Chinese educational leaders know, better than anyone else, what life can be like in the absence of a liberal education. For that was largely the history of China's twentieth century. But conceptions of general and liberal education have limits in a one-Party, Leninist state that is far from guaranteeing the *Lehr-* and *Lernfreiheit* that are central to a Humboldtian enterprise. This leads to a final challenge: can world-class education in the liberal arts and sciences exist in a politically illiberal system? Perhaps, but perhaps only if they are largely self-governed. German universities in the nineteenth century had many political pressures, but they were the envy of the world in part because they also had traditions of institutional autonomy that fostered and (at times) protected creative thinkers.

China's universities today boast superb scholars and among the world's best students. But these students are also forced to sit through required courses in party ideology, and they learn a simplified book version of the history of their own country. Despite new programs of general education, in the realm of politics and history, the distance between what students have to learn in order to graduate, and what they know to be true, grows greater every year. This is a recipe for two types of graduates: cynics and opportunists. In 2014–2015, there were signs that the Chinese Communist Party sought to limit recent trends toward a more liberal education. The reformist president of Peking University who sought to establish an international Yenching Academy, named for Yenching University, was summarily dismissed. The Communist Party Secretary of Peking University, Zhu Shanlu, stressed by contrast the need for enhanced "propaganda and ideology work" at Chinese universities. As Zhu wrote in February 2015:

> Universities are an important battleground for the production and conflu- ence of ideology, and have an important role as leaders, models, and spread- ers of ideas to all of society… For questions of political principles, you must have a resolute standpoint, have a clear-cut stance, dare to catch and dare to manage, have the courage to reveal your sword, and master the important principle of "Academic Research has no boundaries, classroom lecturing has discipline" (*xueshu yanjiu wujinqu, ketang jiaoxue you jilv* 学术研究无禁 区、课堂教学有纪律), and seriously handle public attacks of party leader- ship, attacks on the socialist system, misrepresentations of party and national history, and words and actions that start rumors and create trouble.

Sadly, the greatest obstacle to the emergence of an education in the liberal arts and sciences in contemporary China is the Chinese Communist Party and its political insecurity.

Yet Chinese institutions still are part of the international discourse on general and liberal education, which this volume broadly illustrates through various arguments for a liberal arts and sciences approach to undergradu- ate education in the global twenty-first century. The epistemological argu- ments in favor of focusing on cross- or interdisciplinary themes and big questions; the economic and utilitarian arguments requiring graduates to be equipped with "twenty-first-century skills" for employability and inno- vation; and the social—moral arguments underlining the importance of educating the whole person, social responsibility and democratic citizen- ship (van der Wende 2013b).

The various chapters together confirm that the first two are driven across the continents into a converging global knowledge economy agenda for undergraduate education in the twenty-first century. This may explain the rising popularity of liberal arts and sciences education among employers looking for much sought-after twenty-first-century skills. But they also acknowledge the importance of liberal arts in crafting a public response to the problems of pluralism, fear, and suspicion that their societies face. This argues for an approach to learning that goes beyond utilitarian goals of studying for employment into the development of moral character, inter- cultural understanding, and responsible global citizenship (Nussbaum 2010, p. 125). And as argued by Penprase in Chap. 5, these twenty-first-century skills may not all be so new but could rather be seen as virtues that the complexities of life in this century demand more than ever. They can therefore not be singled out as technical and economic benefits from the political and social institutions that generate innovation, equity, and social cohesion and that are constantly challenged in the global context.

Finally, globalization emerges as the main challenge even for the world's leading universities, as explained by Berkeley's Chancellor Dirks in Chap. 8: "We have only started to come to terms with the volume and velocity of global connections, and have not gone nearly far enough in altering our content and methods to support students in a deeply interdependent world. When planet-wide problems do not recognize either national borders or the boundaries that have traditionally separated academic disciplines, universities must adapt. Any burgeoning university system, too, should take advantage of the opportunity to build around this critical aspect of modern life." To paraphrase Pericles Lewis's chapter title in this book, our mission is to seek liberal education and innovation in Europe, America, and Asia, *and for the World*.

NOTES

1. Most relevant for higher education are the European Commission, Council of the European Union, and the European Parliament.
2. Referring to C.P. Snow's The Rede Lecture on "The Two Cultures and the Scientific Revolution" (Cambridge University, 1959) in which he stated that the breakdown of communication between the "two cultures" of modern society—the sciences and the humanities—was a major hindrance to solving the world's problems.

REFERENCES

Amsterdam University College. (2012). Amsterdam University College liberal arts and sciences for the 21st century: AUC's experiences and achievements 2009–2012. Retrieved from http://www.auc.nl/downloads

Canaves, S. (2015, October 20). Chinese president's speech on the arts: The Hollywood connection. *China Film Insider*. Retrieved from http://chinafilminsider.com/chinese-presidents-speech-on-the-arts-the-hollywood-connection/

Dijkgraaf, R. H. (2009). Liberal arts and the sciences. In Amsterdam University College Impressions of the Grand Opening 22 September 2009. Retrieved from http://www.auc.nl/downloads

Hernandez, J. (2015, October 21). China turns to online courses, and Mao, in pursuit of soft power. *The New York Times*. Retrieved from http://www.nytimes.com/2015/10/22/world/asia/mao-zedong-chinese-history-course-edx.html?_r=1

Miyazaki, I. (1976). *China's examination hell* (trans: Schirokauer, C.). New York/Tokyo: Weatherhill.

Nowotny, H. (2012). *Looking forward: Why you are here and what liberal arts education has to offer.* Keynote Address at the Opening Ceremony for AUC's New Building, Amsterdam, 21 September 2012. Retrieved from http://www.auc.nl/news-events/press-releases/press-releases-kopie/press-releases-kopie/content/folder/2012/09/opening-new-auc-building---dies-natalis-2012.html

Nussbaum, M. (2010). *Not for profit: Why democracy needs the humanities*. Princeton: Princeton University Press.

University Colleges Deans Network. (2014). Statement on the role, characteristics, and cooperation of liberal arts and sciences colleges in the Netherlands. Retrieved from https://www.universitycolleges.info/

van der Wende, M. C. (2011). The emergence of liberal arts and sciences education in Europe: A comparative perspective. *Higher Education Policy, 24*, 233–253.

van der Wende, M. C. (2013a). Amsterdam University College: An excellence initiative in liberal arts and science education. In Q. Wang, Y. Cheng, & N. Cai Liu (Eds.), *Building world-class universities. Different approaches to a common goal* (pp. 89–103). Sense Publishers.

van der Wende, M. C. (2013b). Trends towards global excellence in undergraduate education: Taking the liberal arts experience into the 21st century. *International Journal of Chinese Education, 2*, 289–307.

Wen-hsin, Y. (1990). *The alienated academy: Culture and politics in Republican China 1919–1937*. Cambridge, MA: Harvard University Press.

Wu, J., Yang, X., & Yang, G. (1992). *The scholars*. New York: Columbia University Press.

Zhu, Shanlu. (2015, February 3). *Yi peiyu he hongyang shehui zhuyi hexin jiazhiguan wei yinling zhashi zhuahao xinxingshixia gaoxiao xuanchuan sixiang gongzuo* [Use nurturing and promoting the core values of socialism to show the way; Firmly grasp the university propaganda and ideology work under new circumstances] (Trans: Jocelyn Eby). *Renmin Wang* Retrieved from http://edu.people.com.cn/n/2015/0203/c1053-26497898.html

CHAPTER 2

China's Search for Its Liberal Arts and Sciences Model

Gerard Postiglione

Abstract The liberal arts and sciences have become more important to China as it diversifies its higher education system, builds world-class universities, strengthens its knowledge economy, and takes a greater role in international higher education. Yet, there is increasing debate about how China's indigenous cultural tradition should shape a distinct Chinese model of higher education. Against this backdrop, the chapter reviews selected aspects of liberal arts and sciences in China's top-tier universities, Sino-foreign campuses, and universities in China's Hong Kong Special Administrative Region. The chapter argues that China's challenge in popularizing liberal arts and sciences will hinge the extent to which it addresses rising domestic demands by different groups for higher education with growing aspirations of its universities to go global.

Keywords Universities • Liberal arts • China model • Sino-foreign-campuses

Not since before the 1949 Revolution that brought the Communist Party to power in China has there been as much interest in the Western tradition

G. Postiglione (✉)
The University of Hong Kong, Pok Fu Lam, Hong Kong

© The Editor(s) (if applicable) and The Author(s) 2016   17
W.C. Kirby, M.C. van der Wende (eds.), *Experiences in Liberal Arts and Science Education from America, Europe, and Asia*,
DOI 10.1057/978-1-349-94892-5_2

of liberal arts higher education, not only as a subject of research but as a way to drive science progress in innovative ways. When visiting universities in the early 1980s as a young American professor from the University of Hong Kong, I was struck by the ambivalence toward Western ideas and the factional struggles in the scientific community. The official view still rested on a dogmatic critique of higher education under capitalism. Yet Chinese leader Deng Xiaoping became adamant about having a thousand talented scientists who are recognized around the world. Vogel (2011: 321-22) recounts the story of a 1978 phone call from China to President Jimmy Carter at 3:00AM, Washington time, by his science advisor because Deng Xiaoping wanted approval to send several hundred Chinese immediately to study at American universities, followed by thousands within a few years.

Deng's attraction to American universities was not about liberal arts higher education. It was about science and technology. Deng knew that China's economic reforms and opening to the outside world necessitated preparing world-class scientists. Yet, he also knew it was important to gain a deeper understanding of how other societies made their economies grow. Initially, most Chinese were sent overseas with state financial support to study science and technology. As China's economy grew in the 1990s, the government permitted students to pay their own way as overseas study expanded to encompass other fields including economics, management science, and many other scientific and professional fields.

Twenty years after Deng's message to Carter, I attended an address by President Jiang Zemin in 1998 at the Great Hall of the People in which he declared China's aim to establish world-class universities. This meant an identification of top-tier institutions and a generous package of financial support for strengthening their infrastructure and academic capacity. Shortly after as a senior consultant in 2000 to the Beijing Office of the Ford Foundation, I witnessed the massive investment in university infrastructure, consolidated institutions and expanded enrollments, overhauled science and technology management, efforts to reform university teaching, introduction of more foreign textbooks, and a call to redress the preference for science over liberal arts (Li 2005).

Let us fast-forward to 2010, and the realization that the transition from an economy reliant on labor-intensive export manufacturing to one based on high-tech service would require a greater emphasis on soft skills. The ability of American universities to attract and produce scientific breakthroughs and technological advances has long been attributed to an emphasis on creativity, innovation, and entrepreneurialism. Such achievements in American universities would have been impossible without institutional autonomy and academic freedom. While it is not in the nature of the Chinese university system to set this as the highest priority, there is clearly an acknowledgement that significant reform is needed. Both former Premiers Zhu Rongji and Wen Jiabao acknowledge that China's economic competiveness depend on fostering more creative, independent thinking (Chan 2011; Jiangtao 2011). Qian Xuesen, the father of Chinese rocket science, saw universities as failing to encourage creativity, multidisciplinary breadth, and innovative thinking: "none of our institutions of higher learning is running in the right direction of cultivating excellent talent and is innovative enough" (Zhao and Hao 2010). I heard the same sentiment from the business community when I chaired a discussion with Alibaba's Jack Ma in Hong Kong for the Clinton Global Initiative. Lee Kai-fu, former head of Google China, put it this way: "if you take a college student and drop him into a start-up, there are so many errors he could make, whereas people in the US, they are more independent thinkers who are able to solve problems on the fly and are more suitable as entrepreneurs" (SCMP 2013).

Views gradually began to change about the potential of liberal arts and sciences in Chinese higher education. Overseas Chinese scientists began to play a pivotal role. Physicist Woo Chia-wei, the first Chinese president of a major American university, pointed out the critical importance of liberal arts in American undergraduate education. Woo also became the first president of the Hong Kong University of Science and Technology, a catalyst in the transition of Hong Kong's British-style universities away from a highly specialized three-year program of study to one that extended the bachelor degree programs by adding an extra year with a common core curriculum of liberal arts and sciences for all students, regardless of their specialization. Universities on the Chinese mainland took an interest in Hong Kong's liberal arts initiative as an example of the Western liberal arts integrated into universities in a Chinese society.

This heightened interest in liberal arts and sciences higher education occurred as part of an intensifying debate in China about models of higher education. There was no question, even among many in the younger generation that studied overseas, about the need for a "Chinese" model of higher education, one that addresses fundamental principles, in the same way that fundamental tenets of Western civilization are embedded in the liberal arts and sciences. As the argument goes, China had its own, distinct civilization, and it was not Western. A Chinese scholar and college president reacted to Hong Kong's unapologetic Western model of higher education: "Will Asia be just producing more of the same of the Western-originated contemporary higher education model, or will it be able to unleash a more critical understanding and practice of higher education, a cultural and epistemological reflection of the role of universities as venues of higher learning?" (Cheung 2012, p. 186).

TOWARD A CHINA MODEL

While durable models of education have often evolved organically, from the bottom-up, the history of education in China is marked equally by top-down initiatives. Private academies of classical learning (*shuyuan* 书院) that began in the Tang and continued into the late Qing Dynasty could coexist with a state-led examination system. Today, it may seem contradictory that the current government has established hundreds of Confucian Institutes around the world while it generously funds national research institutes on Marxism. As China returns to the status of a great power, there will probably be more support for models that rely on classical foundations of Chinese civilization. It is notable that Chinese president Xi Jinping took the opportunity to meet Tang Yijie, a leading scholar on Confucianism, on his visit to Peking University. It is also notable that Peking University will be the first to name a building after Karl Marx.

Discussion about a Chinese model of higher education would certainly emphasize the ideas of Cai Yuanpei, Peking University's first president, who sought a synthesis of Chinese and Western thinking. Cai understood the German, French, and English traditions of higher education and considered how these could be brought together with the spirit of Confucianism and other historical traditions (Chen 2015). Others from the modern era, a time of vibrant debate in Chinese education, are also viewed as source for a Chinese model of higher learning, including Hu Shi, Liang Shuming, Ye Yangchu, Mei Yiqi, Jiang Bailing, Yan Fu, Tao Xingzhi, Pan Guangdan, and others (Yang 2003; Hayhoe 2006). Contemporary influential educators would include Gu Mingyuan, Huang Qi, Li Bingde, Lu Jie, Pan Maoyuan, and Wang Fengxian. These thinkers would help arrest the lingering concern that university education has not been sufficiently shaped by indigenous ideas (Yang 2013). There may indeed be good reason for Chinese scholars to look to such thinkers for a Chinese model of a common core curriculum in higher learning.

To sustain the pace of economic growth, China's universities are increasingly expected to play a more powerful role (Postiglione 2011; Kirby 2014). The country has the largest system of higher education, as well as more scientific publications and a larger R&D budget than any other country except the USA (University World News 2007; Jha 2011; Royal Society 2011). It also bodes well for the future that secondary school students in the largest city outperform counterparts in the 60 countries in the Program for International Student Assessment of mathematics and sciences achievement (OECD 2013). The demands on the system of higher education are onerous. They include elevating flagships into the ranks of the world-class universities, uplifting creativity and innovation in science and technology, producing a technically skilled workforce to support an upgrade of production capacity, and addressing issues that could potentially threaten social stability, such as graduate unemployment, unequal access to university due to socio-economic and ethnic disparities. The role of Western liberal arts and sciences education in this set of expectations is increasingly discussed and questioned.

PRECARIOUS BALANCE

China is hamstrung in addressing these expectations by a lack of consensus about how to approach the global movement for liberal studies. It is faced with a precarious balance about how to address rising domestic demands from different sectors for relevant knowledge and skills, status culture, and social stability. Meanwhile it has to find ways to promote internationalization while protecting educational sovereignty, and at the same time, ceding more institutional autonomy to universities. Handling any two is easy enough but juggling all three at once can be a major challenge.

Symptoms of this problem appear in different sectors of higher education. The problem is typified by the increase in overseas campuses in China that cater to the growing middle class but have little impact on the rest of the higher education system. It is typified by the limitations of the South China University of Science and Technology in its experimental effort to be autonomous and offer a reformed governance model. It is typified by the shrinking number of students from poor rural areas in top-tier universities and the growing number of university students who have not secured a job after graduation. With one of the youngest cohorts of academic staff of any system of higher education, and a national demographic profile that will tilt toward the elderly, the time to undertake major reform in higher education is limited.

As China's universities began to move up the global rankings, debate increased about what kind of university model is best for the future (Liu 2010; Kirby 2014).

Therefore, this paper argues that a Chinese model of higher education is not only possible but also necessary if China's university system is to become globally influential in the coming decade. This would be helped along by critical engagement with the Western model of liberal arts and sciences in higher education.

At the very least, China's rise is bringing the global academy toward a better understanding of its historical legacy, one that goes beyond a preoccupation with the imperial examination system, and the accompanying view that it still shapes a style of learning anathema to driving creativity and innovation (Kissinger 2011; Vogel 2011; Hu 2011). Scholarship about historical struggles, developmental experiences, and institutional renovation can also be seen as a creative resistance of Western domination (Jacques 2009; Schell and DeLury 2013; Shambaugh 2013). To avoid this kind of scholarship becoming superficial may require a feisty defense of academic freedom and institutional autonomy. Otherwise, it risks floundering as official lip service to nationalism without a critically independent bite.

Brain Race for Innovative Thinkers

Despite having the world's largest economy, there is widespread concern that Chinese universities produce fewer independent thinkers than their international competitors (Abrami et al. 2014). As the advantage of low labor costs declines, maintaining the country's economic ascent depends on boosting innovation and the quality of human resources through the university system. Generating new products and services will require universities to foster creative and innovative talent, in addition to carrying out cutting-edge research. China's higher education system has expanded to widen student access, but the key challenges are the reform of university governance, enlivening academic culture, more rigorous assessment of quality, and better alignment of university teaching with the changing workplace.

By 2010, about 30% of 18- to 22-year-olds—roughly 30m students—were enrolled in 2263 colleges and universities, including 1079 universities and 1184 higher vocational and junior colleges. Expansion would be untenable without private (*minban* 民办, or popularly run) higher education. Hundreds of private colleges were established to meet the growing demand. Independent colleges run by regular colleges and universities also absorbed some of the demand for higher education. Under the so-called 211 and 985 projects, the government invested heavily in the elite institutions with the aim of creating internationally competitive universities. The 211 project provides extra financial support for 112 universities selected to spearhead national economic development, while the 985 project aims to transform 40 top institutions into world-class universities. Flagship institutions—such as Peking and Tsinghua in Beijing, and Fudan and Jiaotong in Shanghai—have begun to jockey for position in world university rankings. Chinese universities have climbed the league tables by massively boosting their presence in global scientific publications.

Spending on higher education has surpassed US$100 billion. Most of the funding is funneled to the elite institutions, which helps explain why the number of Chinese universities in the Jiaotong global 500 increased from 14 in 2003 to 22 in 2010. Yet a few world-class flagship universities alone cannot carry the whole system of higher education, where the vast majority of colleges trail far behind the front-runners. And even high numbers of scientific publications mask the low frequency with which the world's scientists cite China's scientific publications (4% of the time compared to 30% for the USA). By the end of the first decade of the twenty-first century, China's universities become unrecognizable from their former selves. Yet, they still struggle to deal with long-standing problems such as academic corruption.

Experimentation and the Liberal Arts

Not surprisingly, Beijing's two flagship universities, where most Deans have either overseas degrees or have studied overseas, have taken the lead in introducing innovative practices. Peking University's Yuanpei College is an experiment with liberal arts education, modeled on Harvard's undergraduate general education curriculum. It aims to foster creativity, multidisciplinary thinking, and leadership. Stanford University has established a center at Peking University to expand its role in global education and research, including social science programs that share ideas about social change.

Tsinghua University has intensified the degree of student engagement in learning by introducing classes in group problem solving as well as improving the quality of communication between students and teachers (Hennock 2010). The Schwarzman Scholars Program at Tsinghua University brings future Chinese leaders together with counterparts from all over the world as part of a master degree program. Tsinghua's newest initiative is the establishment of a campus in Seattle. Branded as the Global Innovation Exchange, it will run master degree programs that aim to foster research on new technologies.

Other Chinese universities are experimenting with models of learning that break away from the lecture, textbook, memorization, and exam cycle that is still so common in many universities. Below the top tier of 985 and 211 universities, most colleges and universities still operate with fewer resources, less-qualified academic staff, and less attention from the central government.

The National Outline for Medium and Long-term Education Reform and Development (2010–2020) set a higher education enrollment rate of 40 % by 2020. By then, more Chinese will have a college education than the entire workforce of the USA (MOE 2011). However, China's demographic profile means it has a decade to foster talent among the shrinking proportion of youth who will have to support an increasingly aging population.

International Cooperation amid a Concern about Sovereignty

As China's leadership around the world has grown, its university system has become increasingly engaged internationally. Hundreds of Sino-foreign joint ventures in higher education on Chinese soil were approved. Hundreds of Confucian Institutes for the study of Chinese language and culture were established by the Chinese government on foreign soil. There are plans to establish two Chinese university campuses overseas, one in Seattle, USA, and one in Malaysia. The number of international students coming to China continues to rise, and the number of Chinese self-funded students leaving for overseas continues to grow. Many who go overseas to study do not return, though the number of returnees is on the rise as China's economy opens new job opportunities.

By 2013, there were 1060 approved Sino-foreign joint ventures in higher education with 450,000 students involved. Since 2003, there have been 1,050,000 from higher education institutions (Lin 2014). Sino-foreign cooperation in higher education comes with stern warning about risks to Chinese sovereignty, as a minister of education remarked: "Tough tasks lie ahead for China to safeguard its educational sovereignty as it involves our fundamental political, cultural, and economic interests and every sovereign nation must protect them from being harmed" (Chen 2002). Thus, the debate about liberal studies and sciences education is inseparable from the debate about the establishment of universities with Chinese characteristics. The issue remains embedded within an unambiguous paradox, namely the seeming incompatibility of three elements within its university system: internationalization, institutional autonomy, and educational sovereignty.

The 2003 law on educational joint ventures opened the floodgates to hundreds of partnerships between Chinese and foreign universities. Reforms are underway at top Chinese universities to adapt and innovate on models of liberal higher education customary abroad. Attention is building about whether foreign-partnership campuses can have a significant impact on China's current higher education system. These collaborations and partnerships constitute one type of laboratory for innovative formats in higher learning. While the jury remains out on long-term sustainability of cross-border campuses, both host and guest universities will learn a great deal from cooperation in the running of partnered colleges and universities (Wildavsky 2012).

The majority of international university programs are taught and run by foreign academics, at a substantial premium, within Chinese universities (Postiglione 2009). They are popular with middle-class parents because they give their children the cachet of a foreign education without the cost of studying abroad. In a few cases, foreign universities have gone one step further and set up full campuses with Chinese universities. Nottingham University has a campus in Ningbo; Shanghai Jiaotong and the University of Michigan run an engineering institute in Shanghai; and Xi'an Jiaotong and Liverpool University have established an independent university in Suzhou, among others. In 2013, New York University, which already has overseas study programs in ten countries, opened a new campus in Shanghai with East China Normal University. It will conduct integrated classes in humanities and social sciences, with an equal number of Chinese and foreign students. Duke University has also established a campus in Kunshan in partnership with Wuhan University (Redden 2014). Other American universities with similar aspirations include Kean University and University of Montana.

The rise in Sino-foreign joint ventures has led to more discussion about sovereignty in higher education. An influential scholar of Chinese higher education cautions that permitting foreign entities to hold a majority (more than 51%) of institutional ownership can lead to an "infiltration of Western values and cultures at odds with current Chinese circumstances" (Pan 2009, p. 90). The Vice-Director of Shanghai Education Commission, Zhang Minxuan, makes it clear that a Sino-foreign venture in running an educational institute has to "make sure China's sovereignty and public interests are not harmed" (Zhang, M.X. 2009, p. 33). To do so, at least half of its board of directors have to be Chinese citizens. Zhang Li of the Ministry of Education points out that China's commitment to providing access to its educational market is larger than any other developing country and therefore, "we must safeguard China's educational sovereignty, protect national security, and guide such programs in the right direction"(Zhang, L. 2009, p. 19). Nevertheless, foreign campuses have been had an increased amount of autonomy with less interference from the host campuses since the 2003 law on Sino-foreign cooperation. They must still adhere to regulations set out by provincial-level education bureaus who exert substantial control over student admission and financial issues.

ONE COUNTRY AND TWO UNIVERSITY SYSTEMS

It is in no small way notable that China's Hong Kong, which had only two universities until the 1990s, has three universities that rank in the top ten in Asia, and worldwide has three within the top 50 and five in the top 100 (QS 2012). Hong Kong achieves this with a mere 0.7% of GDP for R&D, placing it about 50th in the world. Such a ranking is due largely to a system of autonomous universities and the grit of their academic staff who enjoy a high degree of academic freedom in a region where supply is limited. Among the reasons cited by Altbach and Postiglione (2012) are steering and autonomy, effective governance, English language in teaching and research, internationalism, the academic professions, and university leadership.

In 2012, Hong Kong universities added a foundation year and a common core of liberal arts and sciences courses for all students. Composed of four areas of enquiry (Scientific Technology and Literacy, Humanities, Global Issues, and China: Culture, State, and Society), the common core aims to develop critical thinking, ability to tackle novel situations, communication and collaboration skills, intercultural understanding and global citizenship, and leadership and advocacy for improvement of the human condition.

The aim of the common core of the University of Hong Kong is to enhance creative and critical thinking; and addressing complex questions of the contemporary world. The overarching goals are: to enable students to develop a broader perspective and a critical understanding of the complex connections between issues in their everyday lives; to cultivate students' ability to navigate the similarities and differences between their own and other cultures; to enable students to more fully participate as individuals and citizens in global, regional, and local communities; and to enable the intellectual, collaborative, and communication skills that will be further enhanced in students' disciplinary studies, and, in turn, contribute to the quality of their lives after graduation. (Hong Kong University 2015)

To raise students' levels of scientific and technological literacy, the common core enables them to engage critically with knowledge and discourse on science and technology, and to respond actively and appropriately to issues surrounding scientific and technological advancements. The specific objectives are: to equip students with a general understanding of the fundamental ideas, principles and theories of science and technology and of natural phenomena and the ways in which scientific and technological knowledge is generated, validated and disseminated, and to enable students to use this knowledge appropriately and effectively; to enable students to understand the form, structure and purpose of scientific language, to read and interpret scientific data and scientific arguments, and at a general level, to evaluate their validity and reliability or claim to knowledge; to arouse students' general interest in science and technology, and to inculcate a willingness and capacity to update and acquire new scientific and technological knowledge; to enhance students' awareness of the circumstances surrounding the history and development of some of the "big ideas" of science, and the social implications of important technologies; to enable students to be critically aware of contemporary socio-scientific and technology issues at the local, regional, national and global levels; to develop students' appreciation of the complexity of inter-relationships among science, technology, society and environment, and the role played by science and technology in the progress of civilization; to raise students' awareness of the moral-ethical issues associated with scientific and technology research and the deployment of scientific knowledge and technological innovations, and to enable them to engage actively with these issues in an ethically appropriate manner; and to enable students to see the interconnection between the humanities and the sciences and technology. (Hong Kong University 2015)

All universities on the Chinese mainland would be hard pressed to institute a Hong Kong-style system when only about 15% of academics in higher education have a doctorate. Thus, while universities on the Chinese mainland search for an indigenous model that is significantly different from the Western one, it may be difficult at the current stage of development. This is not to say that the academic profession on the Chinese mainland would be willing to embrace a Hong Kong model of liberal arts and sciences education.

The Hong Kong case has special relevance due to its national proximity, high number of Mainland born academics, and increasing dependence on the Chinese mainland for research funding as China continues to rise. Hong Kong's universities are in a very different position with regard to the cooperation with universities on the Chinese mainland. Aside from the fact that the Mainland outperforms Hong Kong in the ability of its universities to source funds for innovation from industry, Hong Kong cannot survive without deeper engagement with the Chinese mainland, which is far less of an issue for other countries who are setting up universities in China. If their Sino-foreign cooperation in the running of a university does not work out, a foreign university can step back without consideration as to how it could affect its future economic and political development of its home country. In short, Hong Kong has more at stake. Nevertheless, Hong Kong has managed thus far to successfully juggle a high degree of university autonomy, a high degree of internationalization, and the preservation of Chinese sovereignty as set out in the Basic Law of the Hong Kong Administrative Region of the PRC.

China's strategic plan for 2020 has made the "de-administration of universities" a major objective in raising the academic quality of its universities. This would mean that the government would take more of a steering role than a direct interventionist role in the academic life of universities (Liu 2010). At the same time, the steady and unmistakable rise in the internationalism of China's research universities and the surge in the amount of Sino-foreign cooperation in higher education, including overseas campuses on Chinese soil, and Chinese campuses on foreign soil, are clear indications of reform and opening to international cooperation and experimentation with liberal arts curriculum.

Acknowledgment Hong Kong Research Grants Council HKU37600514, HKU7021-PPR-12

REFERENCES

Abrami, R., Kirby, W., & McFarlan, F. W. (2014). *Can China lead?: Reaching the limits of power and growth*. Cambridge: Harvard Business Review Press.

Altbach, P. G., & Postiglone, G. A. (2012). Hong Kong's academic advantage. *Peking University Education Review, 169-172*.

Chan, M. (2011, April 29). Wen in renewed plea for wider political reforms. *South China Morning Post*. Retrieved from http://www.scmp.com/article/966328/wen-renewed-plea-wider-political-reforms

Chen, H.J. (2015). *Die deutsche Klassische Universitätsidee und ihre Rezeption in China (Deguo Gudian Daxueguan ji qi dui Zhongguo de Yingxiang)*. 3rd edition. Beijing: Peking University Press.

Chen, Z. L. (2002). The impact of WTO on China's educational enterprise and related policies. *People's Education, 3*, 4–7.

Cheung, B. L. (2012). Higher education in Asia: Challenges from and contributions to globalization. *International Journal of Chinese Education, 1*, 177–195.

Hayhoe (2006). *Portraits of Influential Chinese Educators*. Hong Kong: Comparative Education research Centre and Springer.

Hennock, M. (2010, August 9). With new survey, Chinese colleges ask students what they really think. *Chronicle of Higher Education*. Retrieved from http://chronicle.com/article/With-New-Survey-Chinese/123858/

Hong Kong University. (2015). Common core. *Student University Handbook 2015–2016*. Retrieved from http://commoncore.hku.hk/files/CC2015-low-p.pdf?150720

Hu, A. G. (2011). *China in 2020: A new type of superpower*. New York: Harper-Collins.

Jacques, M. (2009). *When China rules the world: The rise of the middle kingdom and the end of the Western world*. London: Allen Lane.

Jha, A. (2011, March 28). China poised to overhaul US as biggest publisher of scientific papers. *Science*. Retrieved from http://www.theguardian.com/science/2011/mar/28/china-us-publisher-scientific-papers

Jiangtao, S. (2011, April 23). Zhu Rongji resurfaces to criticize education reforms. *South China Morning Post*. Retrieved from http://www.scmp.com/article/965804/zhu-rongji-resurfaces-criticise-education-reforms

Kirby, W. C. (2014). The Chinese century? The challenges of higher education. *Daedalus, 143*(2), 145–156.

Kissinger, H. (2011). *On China*. London: Allen Lane.

Li, L. (2005). *Education for 1.3 Billion*. Hong Kong: Pearson.

Lin, J. (2014) Sino-foreign cooperation in the running of schools. Accessed from Sino-foreign cooperation in education: http://www.cfce.cn/a/research/crsrc/2015/0519/2850.html. 2016 June 9.

Liu, D. Y. (2010). *Zhongguo gaoxiao zhica* [The shame of Chinese higher education]. Wuhan: Hubei People's Press.

Ministry of Education. (2011). China's new national plan for medium- and long-term education reform and development (2010–2020). Retrieved June 30, 2014, from http://www.moe.edu.cn/publicfiles/business/htmlfiles/moe/s3501/index.html

Organization for Economic Co-operation and Development. (2013). Asian countries top OECD's latest PISA survey on state of global education. Retrieved June 30, 2014, from http://www.oecd.org/education/asian-countries-top-oecd-s-latest-pisa-survey-on-state-of-global-education.htm

Pan, M. Y. (2009). An analytical differentiation of the relationship between education sovereignty and education rights. *Chinese Education and Society, 42*(4), 88–96.

Postiglione, G. A. (2009). Introduction. China's international partnerships and cross-border cooperation. *Chinese Education and Society, 2*(4), 3–10.

Postiglione, G. A. (2011). Higher education: University challenge. *China Economic Quarterly, 15*(2), 22–25.

QS. (2012). Top universities worldwide. Retrieved from http://www.topuniversities.com/university-rankings/world-university-rankings/2012

Redden, E. (2014, March 12). Bucking the branch campus. *Inside Higher Education.* Retrieved from http://www.insidehighered.com/news/2014/03/12/amid-branch-campus-building-boom-some-universities-reject-model#ixzz34fcVHp8L

Royal Society. (2011). New countries emerge as major players in scientific world. Retrieved from https://royalsociety.org/news/2011/new-science-countries/

Schell, O., & DeLury, J. (2013). *Wealth and power: China's long march to the twenty-first century.* London: Little Brown.

SCMP South China Morning Post. (2013) Former Google China head Lee Kai-fu sows seeds of change, August 8. Retrieved 30 June 2014. http://www.scmp.com/news/china/article/1295061/lee-kai-fu-nurtures-start-ups-and-hopes-better-china

Shambaugh, D. (2013). *China goes global: The partial power.* Oxford: Oxford University Press.

University World News. (2007, November 25). UK: China the next higher education superpower. Retrieved from http://www.universityworldnews.com/article.php?story=20071123120347861

Vogel, E. F. (2011). *Deng Xiaoping and the Transformation of China.* Cambridge: Harvard University Press.

Wildavsky, B. (2012). *The great brain race: How global universities are reshaping the world.* Princeton: Princeton University Press.

Yang, D. P. (Ed.). (2003). *Daxue jingsheng* [The spirit of higher education]. Beijing: Wenhui Publishers.

Yang, R. (2013). Indigenizing the western concept of the university: Chinese experience. *Asia Pacific Education Review, 14*(1), 85–92.

Zhang, L. (2009). Policy direction and development trends for Sino-foreign partnership schools. *Chinese Education and Society, 42*(4), 11–22.

Zhang, M. X. (2009). New era, new policy: Cross-border education and Sino-foreign cooperation in running schools in the eyes of a fence-sitter. *Chinese Education and Society, 42*(4), 23–40.

Zhao, L. T., & Hao, J. J. (2010). China's higher education reform: What has not been changed? *East Asian Policy, 2*(4). Retrieved from http://www.eai.nus.edu.sg/Vol2No4_ZhaoLitao&ZhuJinjing.pdf

CHAPTER 3

The Significance and Practice of General Education in China: The Case of Tsinghua University

Li Cao

Abstract Inspired by the philosophy of liberal education with its origins in both Chinese and Western cultures, a number of Chinese universities have been implementing general education or Suzhi (素质) education with the conviction that all undergraduate students should be broadly educated while trained in a specialized area. This chapter discusses various concepts and the significance and practice of general education in China through a historical examination of why general education has become one of the primary foci of contemporary university reform. Tsinghua University is showcased as an example of how general education is implemented in China with Chinese characteristics in spite of the American influence.

Keywords General education • quality education • Suzhi education • Chinese higher education • Tsinghua University

In recent years, a number of prestigious universities in China, driven by the goals of achieving excellence and leadership in higher education in an increasingly globalized world, have been engaging in various educational

L. Cao (✉)
Tsinghua University, Beijing, China

reforms and pedagogical innovations. One of the cutting-edge scenarios in reforming Chinese higher education appears to be a remarkable variety and expansion of research interests as well as the implementation of general education in arts and sciences. In spite of American influence, general education in China draws on the ancient Chinese tradition of Confucius's "great learning" and the government's recent policy of "quality education" (*suzhi jiaoyu*素质教育), a Chinese version or concept of general education. This chapter discusses various concepts and the significance and practice of general education in China through a historical examination of why general education has become one of the primary foci of contemporary university reform. Tsinghua University is showcased as an example of how general education is implemented in China with Chinese characteristics.

The Chinese Tradition and Means of General Education

The idea of general education is not exclusively a Western concept. A review of Chinese classics reveals that similar concepts appear in early Chinese thought. *The Book of Changes* (*yijing* 易经) declares in its opening paragraph, "we study astronomy in order to detect time variation; we attend to the humanities in order to enlighten the world" (1989, p. 42). *The Great Learning* (*daxue*大学) states, "the way of great learning lies in illuminating one's virtue, loving the people, and abiding by the highest good" and "those who wish to illuminate virtue should cultivate themselves first before putting the family, the state and the whole world in order" (2005, p. 4). Zhuxi (朱熹), the renowned neo-Confucian scholar in the Southern Song Dynasty, preached five ways of learning: "A gentleman should study extensively, inquire prudently, think carefully, distinguish clearly, and practice earnestly" (2005, p. 4). Learning to cultivate one's character and make ethical commitments to society became the central purpose of education in Chinese culture. The Chinese tradition of academies (*shuyuan*书院) is actually an ancient form of the liberal arts college which trained students through reading classics and integrating the learning of arts and sciences in an interpersonal and dialogical way.

In the first half of the twentieth century, during the period of the Republic, Chinese universities were highly influenced by the USA and Germany. Well-known universities such as Tsinghua University, Peking University, and Nankai University called upon and practiced general education to various degrees. When Tsinghua was founded in 1911, its charter ruled: "the purpose of the school is to foster students as whole persons who will enhance national strength with distinguished ability and ethical commitment" (Tsinghua School 1911). Mei Yiqi (梅贻琦), president of Tsinghua University from 1931 to 1948, was known during his presidency for emphasizing general knowledge as primary and specialized knowledge as secondary (*tongshi weiben, zhuanshi weimo*通识为本, 专识为末). He warned that specialists without training in general education would disturb rather than love the people (Mei 2012, p. 8). Therefore during his presidency, he mandated that general education be implemented during the freshmen year. All students were required to take courses from the natural sciences, humanities, social sciences, and other fields regardless of their preference of concentrations. Likewise, Cai Yuanpei(蔡元培), former president of Peking University, insisted that the practice of holistic education should be the fundamental pursuit of a university. Other famous educators such as Zhang Boling (张伯苓) of Nankai University, Zhu Kezheng (竺可桢) of Zhejiang University, Zhou Gucheng (周谷成) of Fudan University were all advocates of liberal or general education during the leadership of their respective universities. Owing to their visions and efforts, these influential Chinese universities formed their own traditions of liberal education that would become a precious resource several decades later when general education was called back into practice at the university.

In the 1950s, not long after the founding of People's Republic of China, in order to meet the practical needs of technical specialists for the planned economy and industrialization, Chinese higher education followed the Soviet model and did an extensive overhaul of its universities and departments. As a result, a formerly multi-disciplinary university such as Tsinghua was transformed into a polytechnic university—"a cradle for red engineers"—with its humanities and social science programs drastically reduced. Although Jiang Nanxiang (蒋南翔), then president of Tsinghua, advocated the policy of "being red and specialized at the same time," specialized education was emphasized over general education, science and engineering over humanities and social sciences. Owing to the Soviet influence and ideological taboos, courses in humanities and social sciences, especially those in Western philosophy and literature, faded out from the university curriculum. While compartmentalization of knowledge made it "convenient" to teach and train engineers and technicians, it deprived students the chance to learn and understand the profoundness and complexity of Chinese and Western thought and culture. The disadvantage of compartmentalized education revealed itself as inadequate and narrow when after 1978 China reopened its door to reform and development. Engineers and technicians who did not adequately understand the relationship between their subjects of learning and the larger society and culture in which they were situated often found themselves disadvantaged at making sense of entangled social issues. This realization became more and more poignant with the acceleration of China's reform and transition to becoming an increasingly active member in the international community in the twenty-first century.

With the rapid development of China's economy and the new wave of globalization in recent decades, the state policy of rejuvenating the nation through "quality education" (*suzhi jiaoyu*素质教育) came into existence and was widely disseminated in every sector of education from primary to tertiary. In September 1995, to address the overemphasis on specialized education at the expense of liberal education and developing students' all-round ability, the Ministry of Education convened a national conference at Wuhan's Huazhong Polytechnic University (since renamed Huazhong University of Science and Technology) in order to launch what was called "cultural quality education" (*wenhua suzhi jiaoyu*文化素质教育). Fifty-two universities were selected to form trial bases for its implementation. (Later, in 2005, the second batch of 61 national bases were set up in 104 universities, which totaled 93 bases across 157 universities.) In March 1998, the Ministry of Education issued "Recommendations on Improving the Culture Quality Education" and defined the concept as follows:

> The basic qualities of college students include moral quality, cultural quality, professional quality and physical and psychological quality, among which cultural quality is the foundation. The cultural quality education we advocate is mainly about humanistic quality education (*renwen suzhi jiaoyu* 人文素质教育). By enhancing arts education for students of science and engineering and science education for students of arts respectively, we hope to raise the cultural and aesthetic taste, and humanistic and scientific quality of all college students. (MOE 1998, p. 1)

The importance of general education in arts and sciences for all students of different majors was stressed in the issued document and a more serious commitment to the humanities was called on. The central idea is that higher education should be professional as well as cultural. College students should be trained in the way that would enable them to deliver professional and cultural values to the society with a strong sense of social responsibility and ethical commitment. The word "cultural," which means nurturing and cultivating one's mind and character, bears a similar meaning as that in Leo Strauss's famous remark that college education should be an "education in culture and towards culture" (Strauss 1959).

The promotion of cultural quality education goes hand in hand with the government policy on quality education nationwide. In 1999, the Communist Party of China's Central Committee and the State Council issued "Decision on deepening education reform and promoting quality education" which made quality education a national imperative. *The National Long-term Education Reform and Development Plan (2010–2020)* issued in 2010 further promulgated quality education as a strategic theme in educational reform. The *Plan* stresses:

> quality education is a strategic theme in educational reform and development. The core issue is what to foster and how to foster it. The key points are: offer quality education to all students, promote all-round development, enhance students' sense of responsibility in serving the country and its people, develop their innovative spirit and practical ability in problem solving. (MOE 2010, p. 7)

In 2012, *The Report of the Eighteenth National Congress of the Communist Party* further made clear the necessity to push forward a comprehensive implementation of quality education, while laying special emphasis on moral education as a fundamental task of college education as a whole. The first and foremost task of the university is to "cultivate students' sense of social responsibility, innovative spirit and practical ability" (Hu Jintao 2012, p. 35). Aiming at reforming the mode of specialized education by facilitating a larger intellectual cosmology and a wide spectrum of education in arts and sciences, cultural quality education can be understood as a transformative localization of American general education in a broad sense. Specifically speaking, the Chinese style of general education is to practice innovative ways of more effective or internalizing education with a focus on cultivating students' mind, character, and competence.

One significant consequence of cultural quality education is that it gave rise to the return of general education to Chinese higher education. Gaining momentum since 2000, general education was used to leverage the reorientation and reform of college education at several top universities in China. Peking University implemented general education through a large number of common elective courses around 2000. Fudan University set up Fudan College in 2005 to guarantee general education structurally in the first year. Sun Yat-sen University founded Boya College in 2009, one of the very first liberal arts colleges within a research university in China. Tsinghua University redefined its college education as a broader professional education with a solid foundation in general education in 2001. It then took the lead in establishing a cultural quality education core curriculum in 2006. Other universities such as Southeast University, Nankai University, and Harbin Engineering University set up similar cores to follow the suit. Xinya College (新雅书院), a pilot residential college with general education as its core task was finally established in Tsinghua in 2014 to push general education further to the core of undergraduate program. With Xinya College being founded at Tsinghua, people believed that the last breakthrough of general education into Chinese top universities had been achieved. Along with Yuanpei College in Peking University, Fudan College in Fudan University, Boya College at Sun Yat-sen University, Xinya College of Tsinghua university joined to form the so-called Four Carts of Chinese general education. In fact, both terms "general education" (*tongshi jioayu* 通识教育) and "quality education"(*suzhi jiaoyu* 素质教育) are used in these Chinese universities regardless of the discrepancies in their respective histories and resources. In spite of the nominal disputes between China and the West about these two concepts, the aspiration to reform Chinese college education along the lines of laying a common foundation of knowledge of science and culture for all students disregarding their majors or concentrations is commonly shared and pursued (*see* Cao 2007).

Why has general education become the center of attention in the Chinese University at the present time? The question needs to be answered in light of the timely reform in Chinese higher education, and more importantly, in light of the rise of China in the twenty-first century.

As the world's second largest economy in terms of GDP, China is becoming a nation that is expected to shoulder larger responsibilities in the international community as a leading developing country. The implementation of an innovation-driven development strategy calls for an education system that can foster innovative and responsible young men and women who are able of coping with the new challenges and problems facing China and all of humankind. Global problems and crises such as global terrorism, global warming, the global financial crisis, global poverty, weapons of mass destruction, nuclear proliferation and radiation, the increasing gap between the South and the North all pose challenges that need strong and competent leadership who require larger cross-border perspectives and methods. The four abilities outlined by the Harvard "Red

Book," *General Education in a Free Society* (1945), are not outdated in today's context of promoting general education: "to think effectively, to communicate thought, to make relevant judgments, and to discriminate among values" (p. 64).

In retrospect, the renewal or rise of general education in China corresponds to the international trend of renovating general education in the university. In the first decade of the twenty-first century, world-renowned universities such as Harvard, Columbia, Melbourne, and Tokyo all began to review and reform their respective programs of general education. Harvard completed its *Report of the Task Force on General Education* in 2007, University of California Commission published *General Education in the 21st Century* in the same year. The University of Melbourne adopted "the Melbourne Model" in 2008 whereas Columbia University called on a global core at the same time. Meanwhile, top universities in East Asia such as Tokyo University, Seoul National University, National University of Singapore, and Chinese University of Hong Kong all set up new programs to strengthen liberal or general education. To China, whose dream of a strong nation largely rests upon the quality of education, especially higher education, the implementation of general education and quality education is conceptualized as strategically important and necessary. In order to build China as an innovative country, the Chinese university must train its students to be able to work and live in an increasingly competitive and interconnected world. Meanwhile, China's increasing influence and share of international responsibilities require her younger generation to be able to create new knowledge and value concepts and contribute them to the whole world. If education is a cultural course, Chinese students should be well prepared culturally for today's interdependent and globalized environment. In order to lead society to thrive in the twenty-first century with new ideas and better methods, they should have full awareness of their own cultural roots as well as a good understanding of other cultures. They should be culturally and intellectually competent in delivering cultural values and professional expertise at the same time to the ever-changing world so as to resolve new problems in a new era.

From the perspective of a Chinese multi-disciplinary research university, the stress on general education does not mean abandoning specialized education. On the contrary, a good specialized education is not only necessary but also compelling. As pointed out by the Association of American Colleges and Universities (AACU), "Today, a liberal education usually includes a general education curriculum that provides broad learning in multiple disciplines and ways of knowing, along with more in-depth study in a major" (AACU 2015). This is especially true in a Chinese university. The importance and necessity of combining general education with specialized education and science with liberal arts is widely recognized and has become a theme in current university reform. The idea and belief of integrating the two in an interactive way is closely related to the ancient belief of unification between man and nature (*tianren heyi* 天人合一), humanities and science (*wenli jiehe* 文理结合), knowledge and action (*zhixing heyi* 知行合一). With this belief, we may aim Chinese general education at the following targets:

1. To prepare students to work and live for the development of society as well as of the individual;
2. To lay a broad foundation of knowledge in the arts and sciences for developing intellectual and civic capacities;
3. To establish a consensus of values in the transformative society while making sound use of traditional values of Chinese culture and the core values of socialism;
4. To develop cultural awareness and cross-disciplinary and cross-cultural perspectives in an increasingly challenging and interdependent world of globalization;
5. To train students' ability to relate, synthesize, create, and transfer knowledge;
6. To cultivate a larger sense of responsibility for society and all humankind.

While the vision and purpose are abstractly outlined, the implementation and actualization demands huge effort. As reforms in higher education by means of implementing general education rapidly take shape in a number of leading universities in China, general education in each university achieves its own characteristics and style. The case of Tsinghua is only one of many examples.

THE PRACTICE OF GENERAL EDUCATION AT TSINGHUA UNIVERSITY

Starting in 1977 when college enrollment by national entrance examinations was recovered for the first time since the Cultural Revolution (1966–1976), general education at Tsinghua went through several stages. The first stage mainly saw the emergence of elective courses, which took place in the 1980s and 1990s. In 1981, elective courses in humanities and social sciences were introduced in addition to the required courses in Marxism and Mao Zedong Thought. Six years later, in 1987, five credits in humanities and social sciences courses were required of all undergraduates. In 1995, a trial base was established for cultural quality education when elective courses in Chinese and Western classics were taught in a way that cared for both text and context. The second stage began in 2001, when 13 credits of Culture Quality Education (mainly courses in humanities and social sciences were prescribed in the undergraduate program, which made 39 the total required credits in general education: 14 in Political Studies, 8 in College English, 4 in Physical Education), which was equal to 1/4 of the all undergraduate education credits. This breakthrough in the credit system finally led to the introduction into the curriculum of Core Courses of Cultural Quality Education (*wenhua suzhi jiaoyu hexin kecheng*文化素质教育核心课程) in six categories in 2006: Philosophy and Life, History and Culture, Literature and Language, Art and Aesthetics, Science and Society, and China and the World. The introduction of the Core Courses into the prospectus is as significant as the resolution of 13 credits, for it marked a structural platform to further develop Core Courses in general education for the years to come. The Core was designed according to the following principles and goals (Tsinghua University 2007, pp. 70–71):

1. To teach students to profoundly understand Chinese civilization in the general context of world civilization.
2. To enable students to philosophically understand the relationship between man and nature, man and society, man and himself.
3. To help students read and appreciate great works of Chinese literature and world literatures.
4. To equip students with knowledge and understanding of the development and application of new science and technology.
5. To prepare students to meet the challenges of globalization with adequate understanding of the relationship between contemporary China and the world.
6. To develop transferable and practical skills of communication in both oral and written forms.
7. To cultivate students' artistic ability and aesthetic judgment.
8. To help students achieve a wider scope of knowledge, and become able to learn and renew knowledge for life.

The New Humanities Lecture Series, launched in 2005 as a public lecture series, became, in 2008, a compulsory lecture-course one credit from which is required of all undergraduate students. In 2009, about 90 courses of general education in arts and sciences were offered, covering eight categories: Philosophy and Ethics, History and Culture, Literature and Language, Art and Aesthetics, Environment, Science-technology and the Society, Contemporary China and the World, Life and Career Development, and Mathematics and Natural Science. In the same year, the School of Economics and Management, Qian Xuesen Class of Mechanic Engineering (*qianxuesen lixue ban*钱学森力学班), initiated their own reforms in general education, which ultimately formed a stimulus to the reform and renovation of general education university-wide. In 2011, the Freshman Seminar was merged into the Core and eight credits of the Core were required of all students instead of the four in the preliminary year of 2006.

The first and second stages over the last two to three decades have immense significance both conceptually and practically since they successfully laid a solid foundation for further progress and reform. When the credits and categories of the Core were designed, the quality of the courses themselves remained to be further constructed and developed. This brings us to the present stage of improving and consolidating the general education system while striving to achieve true pedagogic results.

The present stage, the third stage, began in 2014 when the university decided to further strengthen the Core Courses of cultural quality education while initiating a pilot project to further incorporate general education into the undergraduate program. Xinya College, a residential college with general education as its center of focus came into being as a result of reform and uplifting of general education. The author, Cao Li, was appointed director and Gan Yang, dean of Boya College, was appointed general supervisor of Xinya College.

Founded in September 2014, Xinya College took in students from schools and departments which see general education as crucial to undergraduate education. A General Core is designed especially for Xinya students with a sharper focus on Chinese and Western civilizations and the interconnectedness between different civilizations and cultures. Students and professors are expected to learn and teach in the way that brings the ancient and the modern, Chinese and Western, sciences and humanities into mutual illumination.[1]

At present, Xinya takes in students from Architecture, Life Science, Law, Aviation and Aerospace Engineering, Electronic Engineering, and Automobile Engineering. The selection was made through a consensus among disciplinary schools and departments, students themselves, and Xinya College.[2] In order to achieve in-depth study and active interactions among students, all general education Core Courses are designed for and conducted in small-size classes. Students are expected to treat readings and assignments as seriously as those requested in their major courses. They are required to earn 30 credits over the first two years from the General Core of arts and sciences plus 18 credits in political studies and college English. Aside from course sets designed for respective subjects and majors, cross-disciplinary programs such as Philosophy, Politics, and Economics (PPE), Electronic Engineering and Life Science (EELS), and Art Design and Creative Engineering (ADCE) are being designed for students whose capacity and interest goes beyond one single subject. Those who distinguish themselves in both the General Core and major studies or cross-disciplinary programs can apply for an honors degree.

Drawing on the American practice of a liberal arts college within a multi-disciplinary research university with a critical eye and a strong awareness of its own cultural roots and educational tradition, the steering members of Xinya College believe that the Chinese liberal arts college must be rooted in Chinese culture and society. It should develop itself in accordance with the academic tradition of the university while creating a new collegiate system bearing Chinese and Tsinghua characteristics. In fact, Xinya College is born out of the legacy of general education handed down from the old Tsinghua and the accumulation of 20 years of continuous effort in implementing cultural quality education in the new Tsinghua. As pointed out by Chen Jining, then president of the university, at the founding ceremony, "the founding of Xinya College is a strategic plan for the present as well as for the future of the university."[3] Vice-president Yang Bin observed the late-birth of Xinya College as a cautious decision of Tsinghua authority after examining various experiments in general education home and abroad. "Doing it right is more important than doing it early."[4]

China is going through a profound social transformation that is accompanied by unprecedented opportunities and challenges. Given the irreplaceable role that the university plays in shaping a civic society, the implementation of a general education that aims at cultivating students' minds and characters and educating them to live a meaningful life can be China's promise of the future. A common core that takes knowledge, ability, and judgment of values as the central components of a shared cognitive foundation and value system would hopefully lead students onto a promising path of self-fulfillment and social commitment. The Chinese university that has a compelling vision to illuminate intelligence and virtue would commit itself continuously to the well-being of the nation as well as the world. This sense of aspiration and obligation will always remain at the heart of a great university.

NOTES

1. See Appendix for Xinya Core Courses 2014–2016.
2. From 2016 on, Xinya College will accept students directly from high school through entrance examinations without limiting itself to these five schools or departments. Students can declare their majors after they enter college.
3. Remarks made by Chen Jining at the opening ceremony of Xinya College on September 27, 2014.
4. Remarks made by Yang Bin at the annual meeting of Xinya College on January 12, 2015.

APPENDIX A: CORE COURSES FOR XINYA COLLEGE 2014–2016

No.	Course	Instructor	Department
1	Close Reading of *The Records of the Historian*	Prof. Xie Siwei	Dept. of Chinese Literature
2	Early Chinese Civilization	Prof. He Xudong	Dept. of History
3	Law and Literature	Ass. Prof, Zhao Xiaoli	Law School
4	The Inspiration of Art	Prof. Li Mu	Academy of Fine Art
5	Selected Readings in Chinese Modern Literature	Prof. Wang Zhongcheng	Dept. of Chinese Literature
6	Literature and Creative Writing	Prof. Liu Yong	Dept. of Chinese Literature
7	Western Thought and Modern Society	Prof. Huang Yusheng	Dept. of Philosophy
8	19th-Century English Literature and Art	Lecturer Gao Jin	Dept. of Foreign Languages and Literatures
9	Artifacts in Life: Ancient Chinese Craft	Asso. Prof. Chen Yanshu	Academy of Fine Art

10	Science and Civilization	Prof. Liu Bing	Institute of Science, Technology and Society
11	Architecture and City	Prof. Wang Yi	School of Architecture
12	Sovereignty and Human Rights	Assi. Prof. Liu Han	Law School
13	The Cognitive Foundation of Trans-disciplinary Studies	Asso. Prof. Gu Xueyong	Dept. of Industrial Engineering
14	History of Sui Tang and Five Dynasties	Prof. Zhang Guogang	Dept. of History
15	Reading Mencius	Prof. Tang Wenming	Dept. of Philosophy
16	The Universe in Ancient Greek Philosophy	Dr. Liu Weimo	Xinya College
17	New Readings of Ancient Chinese Myth	Lecturer Li Ping	Law School
18	Modernity and Post-Modernity in International Law and Relations: Texts for a Variety of Humanisms	Prof. Tony Carty	Law School
19	Perspectives in Political Science	Asso. Prof. Ou Shujun	Dept. of Political Science, Renmin University
20	The Idea of the University (Freshman Seminar)	Prof. Cao Li	Xinya College
21	The Self, the Other and the Society	Prof. Gan Yang	Xinya College

REFERENCES

Association of American Colleges and Universities. (2015). What is a 21st century liberal education? Retrieved from http://www.aacu.org/leap/what-is-a-liberal-education

Cao, L. (2007). Thoughts on cultural quality education and general education [*Guanyu wenhua suzhi jiaoyu yu tongshi jiaoyu de panzheng sikao*]. *Qinghua daxue jiaoyu yanjiu*, (07)2, 24–33.

General Office of the CPC Central Committee. (1999). Document no. 9. Retrieved from http://www.moe.edu.cn/publicfiles/business/htmlfiles/moe/moe_177/200407/2478.html

Harvard Committee. (1945). *General education in a free society: Report of the Harvard Committee*. Cambridge, MA: Harvard University Press.

Harvard University. (2007). *Report of the task force on general education.* Cambridge, MA: President and Fellows of Harvard College.

Hu, J. T. (2012). *Firmly march on the path of socialism with Chinese characteristics and strive to complete the building of a moderately prosperous society in all respects: The report of the eighteenth National Congress of the Communist Party.* Beijing: People's Press.

Mei, Y. Q. (2012). Da xue yi jie [Explaining the university]. In *Zhongguo de daxue* [China's universities] (pp. 2–12). Beijing: Beijing ligong daxue chubanshe. (Original work published 1941).

Ministry of Education. (1998). *Guanyu jiaqiang daxue xuesheng wenhua suzhi jiaoyu de nuogan yijian* [Recommendations on improving university students' culture quality education]. Retrieved from http://jwc.gxu.edu.cn/images/14/11/13/4bode6ibev/53C90B2F9AD8C9B753EE73D3A1EB80F1.doc

Ministry of Education. (2010). *The National Long-term Education Reform and Development Plan (2010–2020).* Beijing: Zhongguo fazhi chubanshe.

Strauss, L. (1959). An address delivered at the tenth annual graduation exercises of the basic program of liberal education for adults [Presentation transcript]. Retrieved from http://www.ditext.com/strauss/liberal.html

Suyongdian School. (1989). *Yijing.* Beijing: Beijing daxue chubanshe.

Tsinghua School. (1911). *Qinghua xuetang zhangcheng* [Constitution of the Tsinghua school].

Tsinghua University Office of Teaching Affairs. (2007). 2007–2008 *xueniandu benkesheng xuanke shouce* [2007–2008 Academic year undergraduate course selection manual].

University of California Commission on General Education. (2007). *General education in the 21st century.* Berkeley: Center for Studies in Higher Education.

Xi, Zhu. (2005). *Daxue, zhongyong, lunyu* [The great learning, the middle way, the analects]. Maoyi Zhang (Ed.). Xi'an: Sanqin chubanshe.

Xinya College. (2014). Archive of the general education curriculum.

CHAPTER 4

In Asia, For the World: Liberal Education and Innovation

Pericles Lewis

Abstract This chapter explores the ways in which both new and old institutions might leverage the past successes and challenges of liberal education systems in order to create a form of education that emblematizes, promotes, and sustains innovation. Yale-NUS College's common curriculum encompasses both Asian and Western influences in humanistic, social, and scientific studies. Students participate in on-campus communities of learning while expanding the scope of inquiry outwards through research trips and internships. In drawing together a highly international group of students and challenging them to create connections across time, space, and cultures, this form of liberal education teaches students to take risks and experiment so that they in turn may become innovators in the university and in the world.

Keywords innovation • liberal education • Yale-NUS college • common curriculum • Asia

Liberal arts education is among the most honored and the most contested creations of American colleges and universities, a mode of learning broadly and deeply which has inspired new programs and schools throughout Asia and

P. Lewis (✉)
Yale-National University Singapore, Singapore, Singapore

© The Editor(s) (if applicable) and The Author(s) 2016   47
W.C. Kirby, M.C. van der Wende (eds.), *Experiences in Liberal Arts and Science Education from America, Europe, and Asia,*
DOI 10.1057/978-1-349-94892-5_4

beyond, even as it has faced increasing skepticism within the USA. Although liberal education is often understood as a primarily Western or even American invention, it in fact draws on ideals recognized throughout Asia. It is recorded in the Analects of Confucius that "The Master said: "The gentleman [*junzi*] is not a vessel [*qi*]," (*Analects* 2.12). The recent translator Chin Annping explains, "A gentleman, *junzi* (君子), is broad of spirit and intellectually agile; he can take on different problems and apply himself to many situations and so is not a vessel, a *qi* (器), for a specific use." How should we who seek cooperation between Asia and the West educate men and women of broad spirit, intellectual agility, and great virtue today? Three themes dominate recent discussion of what is healthy and what is ailing in the tradition of liberal arts education: conversation, character, and community.

Twentieth-century liberal arts education should draw on the strengths of the tradition. We must at the same time acknowledge certain important criticisms of existing liberal education programs, which question whether and how such programs might successfully mold students' character along with their minds. This chapter draws on the experience of founding a new college of liberal arts and sciences in Singapore, Yale-NUS College, envisioned as a community of learning where living and learning are intertwined and where students and faculty share a common goal of developing an innovative new educational model.

While this chapter speaks broadly of liberal education, my focus will be on the form of education in the liberal arts and sciences practiced in the best colleges and universities in the USA, as contrasted with university systems that emphasize relatively early specialization, such as have been common in Asia and Europe at least since the Second World War. While I myself attended such a program at McGill University, where I took a three-year honors degree in English literature, my experiences in graduate school at Stanford University, as a faculty member at Yale University, and most recently as president of Yale-NUS College, has convinced me that a broader, four-year program spanning the breadth of the humanities, social sciences, and natural sciences promises a better foundation for future global citizens.

In a wider sense, of course, my training at McGill was also a form of liberal education, as a broad study not directly related to professional goals. Perhaps more important than the distinction between curricula with greater specialization or greater depth is the broad spirit of liberal learning fostered by many great colleges and universities around the world, and certainly not the exclusive preserve of the American form of liberal education. The focus of this chapter, then, is dual: both on the specifics of a particular kind of curriculum and collegiate organization, and on the broader principles underlying liberal education in general. The form of liberal education I investigate and advocate here privileges no single field of study over the others, but instead argues for their integration in the education of globally aware citizens.

There are at least five good reasons to pursue a liberal education and to provide one for our students.[1] The most commonly cited reason, and a very important one, is to make students into better-informed citizens. By developing their critical reasoning skills, and by practicing the arts of discussion, collaboration, and compromise both inside and outside the classroom, students become better able to debate matters of public importance and to arrive at a reasoned agreement, or reasoned disagreement, with their peers in the political or civic sphere. A second reason, equally valid and perhaps even more significant to some parents and governments, is to shape more innovative contributors to the economy and society. Technical education is extremely important for the development of industrial society, but in the postindustrial world, employers value "softer" skills such as creativity, the ability to "think outside the box," and openness to multiple perspectives. Liberal education fosters these traits. Third, certain forms of liberal education also prepare students well for life in a multicultural or cosmopolitan society by making them aware of a variety of cultures and the need to communicate effectively across cultures. Fourth, and more fundamental than any of these, perhaps, is the ethical case for liberal education. Socrates said that "the unexamined life is not worth living." Liberal education makes us aware of the importance of examining our own prejudices and assumptions by fostering habits of self-awareness and self-criticism. Finally, and most intangibly, liberal education allows the individual a greater enjoyment of life, whether it is in appreciating a work of art, understanding an argument in philosophy or an equation in mathematics, or exploring the diversity of the natural world (Delbanco 2012, p. 9–35).

The challenge for new programs in liberal arts and sciences throughout Asia is to foster the vibrant *conversations* that have developed within the tradition of liberal arts education, to identify and ameliorate certain shortcomings of contemporary liberal education with regard to student *character*, and to promote the innovations that might reenvision liberal education as the basis for a new intellectual *community* within a complex, increasingly interconnected world.

CONVERSATION

The liberal education of the twenty-first century should draw on the traditional strengths of the liberal arts tradition, which can be understood as a series of conversations: conversations between past and present, conversations between cultures, and especially conversations between students and faculty. The philosopher Hans-Georg Gadamer spoke of the encounter with a text from the past in terms of a "fusion of horizons." As Gadamer explains in an important passage from *Truth and Method*, "The horizon is the range of vision that includes everything that can be seen from a particular vantage point" (2004, p. 317, 313). Metaphorically, we can speak of people with narrow or broad horizons, that is, people who have very limited ideas and people who perceive many other points of view. What is important about the horizon analogy is that it both suggests that we can see a certain distance and calls attention to the limited range of our sight. There is always something beyond our horizon that we do not yet know or have not yet experienced. The nature of conversation, including conversations with texts, is that we try to make our horizon match the horizon of the person to whom we are speaking. Gadamer gives as an example of an unproductive conversation the oral examination, in which the examiner seeks to find out what the examinee knows, but not really to learn from him or her, and not to arrive at a real understanding (Gadamer 2004, p. 314). In contrast, the true "fusion of horizons" for Gadamer consists in really engaging with the other and thus in opening up our horizon for possible changes. Our horizon evolves as we grow, learn, and develop. A real fusion of horizons with the past involves the potential for such change. Our encounter with the past, Gadamer argues, is part of what allows this development. The problem of interpretation is the problem of all understanding—of how to engage in this dialogue with others, with texts, with the world—by which we at once challenge our own horizons and seek to learn from and speak back to the rest of the world.

Gadamer was concerned with the philosophy of interpretation, but it is no accident that the most characteristic form of liberal arts education in the USA is the undergraduate seminar, in which a professor and a group of students grapple with the interpretation of an important text, a work of art, or a piece of evidence. Liberal education allows students to test their own ideas against those of their classmates, their professors, the great works and thinkers of the past, and the most important current research in their fields of study. It also demands that they learn some of the tools of interpretation in a variety of disciplines, so that they can approach problems from multiple perspectives. Ideally, liberal education also leads students from different backgrounds to encounter diverse and disparate cultures. Not only do students bring these distant cultures together through their studies, but they themselves encounter different backgrounds and patterns of thought through the diverse student body with which they share this educational journey. Indeed, liberal education leads students not just to encounter distant cultures, but to redefine what they understand as "distant," forming their own pictures of the world and its global conversations.

Liberal education encourages students to enter into such conversations firsthand, through small classes, discussion groups, and laboratories. Here, they become active participants in their own learning process, deriving new, diverse methods for encountering problems ranging from social inequality to the age of the universe. Students work together to solve problems, broaden their perspectives, make decisions based on others' advice, and learn accountability for their own ideas and contributions.[2] This feature of liberal education is one of the reasons that it can propel students toward innovative approaches, which can be valuable in their later professional lives. By working alongside others, students come to recognize that innovation is the product not just of a single brilliant mind operating in a vacuum, but of the continued, concerted efforts of teams that discuss, encourage, criticize, test, and refine new ideas.

Such conversations between individual students and the group or groups in which they participate become part of daily life through the residential college system. The collegiate model goes all the way back to medieval Oxford and Cambridge, but similar communities of learning existed in China and India even earlier.[3] By living alongside peers with a variety of different backgrounds, experiences, and interests, students learn to coexist with others, even in situations where their opinions or expectations may differ widely from one another (*see* Lewis 2007, p. 79). In such communities, students participate in sports, clubs, societies, musical groups, and student publications, creating a lively civil society in parallel with the official curriculum taught by the faculty. Supported by a residential staff that pays attention to their emotional and social needs, students find in this period of their lives an opportunity for personal growth and for developing their abilities as citizens and leaders. Residential colleges continue the work of liberal education beyond the classroom, promoting compromise over unilateral decision-making and a recognition of others' humanity and worth over the primacy of a single student's individual needs. Students become leaders among their peers, but also learn to listen to what their peers have to say, forging and evaluating solutions together. Further, residential colleges lead students to see their education not as a "job" that is localized to a single classroom or laboratory but as a vocation, a collection of many "roles" that they play in relation to others.[4]

By learning to move fluidly among these roles and to integrate education into their everyday lives, students create a space in which they might reevaluate and adapt the lessons of the classroom into real-world conversations.

CHARACTER

Despite liberal education's broad success in fostering an environment that promotes conversation, there remain aspects of liberal education as currently practiced that call for reform and innovation. Older liberal arts institutions of the West and new or recreated institutions of Asia have both taken on this task. To say that liberal education requires some rethinking in the twenty-first century is simply to recognize that colleges and universities, like other great institutions, must change in response to history, technology, and the needs of our students (*see* Laloux 2014, p. 15). This section outlines a few of liberal education's pitfalls, in order to address how this form of learning might be improved and enlivened to serve the needs of twenty-first-century students. The most telling critiques of liberal education can be summarized in terms of the problem of character. On the one hand, colleges and universities have often promised to shape the character of their students. On the other, critics such as the former Harvard College Dean Harry Lewis and William Deresiewicz (a former colleague of mine at Yale), have pointed to weaknesses, especially in elite American educational institutions, that they claim breed excellence "without a soul" or worse, "excellent sheep" (Deresiewicz 2014). While their analyses may be exaggerated or overly pessimistic, they do indicate real problems. As competition for entry into the top institutions continues to increase, there are risks that those who attend elite institutions such as Ivy League colleges will come from a narrower stratum of society, will see themselves as entitled, will avoid risks and stick to safe subjects and pursuits, and will be treated by their institutions as the customers who are "always right" rather than challenged to grow and sometimes to fail.

The most important issue here is one of access. Given increasingly tough competition for spots in the top universities of the world, those born into privileged families can take advantage of better primary and secondary education, tutors, admissions coaches, and other perquisites that help them to gain admission to elite institutions. In the USA, these institutions have recognized the problem and devoted considerable resources to seeking out students from poor backgrounds or from underrepresented minorities, but it is still the case that most students at Ivy League colleges come from families that are upper middle class or wealthy, and that relatively few talented students from the poorer segments of society receive the kind of college preparation that allows them to attend the Ivy League. In 2000–1, for instance, only about 4.4 % of students at Harvard University came from families making less than US$30,000, and according to a 1995 study, only about 12 % of students who even *applied* to Ivy League schools came from families in the bottom income quartile (Pallais and Turner 2007, p. 130–1; see also Bowen 2005, p. 95*ff.*). Even when underprivileged students do succeed in attending an elite institution, they are sometimes underprepared for the work they find there (*see* Lewis 2007, p. 134). There has been some improvement in the past 15 years at Harvard, Yale, and other elite institutions, and notably also at selective liberal arts colleges such as Amherst and Vassar.

Even among those who might benefit from an Ivy League education, ignorance of the opportunities or fear of the cost may prevent them from applying. This is not a problem with an easy solution; colleges and universities rightly seek to promote diversity, but ultimately some of these problems are a product of a stratified social structure and the poor opportunities at earlier levels of education for the underprivileged in the USA. Elite universities such as Yale and Harvard have made a point in recent years of seeking out a more diverse student body through generous funding (Yale University 2012, p. 3). While we strive to improve public education in the USA, colleges and universities should continue to seek out a diverse array of students with high potential and should place greater emphasis on developing an ethos of service among their students so that those who are privileged to attend the great universities recognize their responsibility for giving back to the broader community. Yale-NUS College has similar values and strategies. In selecting students, we look for not only those with a strong record of academic achievement and desire to become critical thinkers capable of rigorous and insightful analysis, but also individuals with different backgrounds, interests, and ambitions, including an interest to serve broader society, both locally and globally. A number of student organizations reach out to the local population, for example, teaching special classes for at-risk high school students or migrant workers.

There are those who argue, however, that the problem with our admissions systems goes deeper than emphasizing service to others and seeking out students from diverse backgrounds who also have a commitment to humanitarian ideals. Even among the children of the elite who do ultimately get into the top universities, there may be negative effects from the pressure of competing for these spots. One risk is that well-meaning but ambitious parents might teach their children to view their studies and extracurricular activities only through the prism of what will get them into the college of their dreams. Here, it would be helpful if parents and admissions officers reminded students that there are many good colleges and universities where they can get an excellent education and that the disproportionate focus on a very few elite institutions can be damaging and misleading.

One of the main criticisms of the liberal education provided once students arrive at leading institutions has been that it caters more to student desires or fads than to challenging students or building their characters. As Lewis (2007) notes, liberal education systems based on "distribution requirements" provide "the easy way out of the imperative for general education" for both students and faculty, since professors "can teach from their home bases and yet take credit for contributing to the breadth of undergraduate education" and students can "treat curricular requirements as the rules of a game they are challenged to win, seeking out the easiest course in each division" (p. 50). In many curricula, such relatively weak general education requirements have allowed the opening of a rift between disciplines, with the humanities and natural sciences on opposite sides of this divide, and the social sciences hovering somewhat uncertainly toward the middle. In this situation—a modern variant of C.P. Snow's "two cultures"—students and faculty seem increasingly unable to traverse the space between the disciplines.[5] Humanities majors think that science is too "difficult" or "objective"; science majors think that the humanities are too "soft" or "subjective." If given the opportunity, many students avoid taking courses outside of their own discipline, or, in the case of "distribution requirements," they often take the easiest, most watered-down courses they can find. This situation may help students to specialize in a single field, but it also runs the risk that they will remain distrustful of concepts and forms of learning beyond their own purview. In fact, this isolation of the disciplines stands in the way of students bringing together multiple disciplines in order to make all of their studies richer by adapting vocabularies and ideas from one discipline to another in order to create an interdisciplinary conversation.

Furthermore, in many liberal education programs and particularly in the USA, curricula artificially limit the scope of their students' education even before they arrive on campus. Even in institutions that boast cohesive, challenging liberal education curricula, the focus is often exclusively on Western thought, with relatively weak gestures toward comparison with sources outside of Western Europe and Northern America. In addition, many schools place a disproportionate focus on liberal education in the humanities, with students required to take only superficial, watered-down courses in quantitative or natural sciences. This exacerbates the "two cultures" problem, and, ironically, does a disservice to both sides of the disciplinary divide: it paints the humanities as easy enough for any student to learn, marks the sciences as too difficult, and provides students with a rather lopsided, insufficiently challenging education.

Finally, a common complaint about academia today concerns the perceived over-emphasis on research at the expense of teaching. While there are indeed a number of problems with the incentive structure in academia that sometimes grants tenure and promotion to indifferent or even bad teachers, and that discourages some faculty from truly engaging with their students, it would be a mistake to assume that the relationship between research and teaching is a zero-sum game. Some faculty may ignore their teaching duties while pushing to get tenure; others may churn out publications of minimal importance. For the most part, however, the opportunity to conduct specialized research allows faculty to develop their knowledge of their fields and to hone their own intellects, allowing them to share this learning, in turn, with their students. There is, however, some risk that by over-emphasizing research in our rewards system (tenure, promotion, salary), we may encourage faculty to become too absorbed in their research at the expense of their teaching. This phenomenon can also have a negative impact on the content of their teaching, as they may treat undergraduates like proto-graduate students, grading undergraduate work as if it were a credential for graduate school rather than tailoring courses to fit undergraduate needs, and expecting students to become premature specialists rather than allowing them to explore subjects broadly.

The challenge here lies in balance. Faculty should have the freedom to pursue their own research, so that they can advance knowledge and infuse their teaching with the newest ideas and methods. They should also teach a curriculum that is inspiring and demanding, but tailored to a class of undergraduates who likely will not become specialists in a given academic field. Instead, professors should teach students with the expectation that they will be adapting their liberal education to a world beyond the academy, from the arts to law, medicine to business, government to non-governmental organizations. They need not pander to students' career goals but should allow this breadth of application to infuse their approach to their specialized subjects, so that both students and faculty see beyond the subject at hand to its larger importance.

COMMUNITY

One further challenge of current liberal education programs bears mentioning: the problem of forging a deeper link between living and learning, between undergraduate student life and the educational mission of the college or university. As twenty-first-century educators, how might we address these challenges of liberal education while also retaining the traditional strengths of this mode of education? How might we create an innovative form of liberal education which itself promotes change? These are questions that my colleagues and I have asked ourselves repeatedly as we have forged ahead in the process of creating a new college of liberal arts and sciences at Yale-NUS College in Singapore. Yale-NUS is an autonomous college within the National University of Singapore, jointly governed by NUS and Yale University. Its founding has provided a rare and exciting experience to draw on the history of the liberal arts and sciences and on current best practices, and then to apply the resulting findings to a practical outcome.[6] At Yale-NUS, this is precisely the opportunity we have cherished over the past several years. Some of the most innovative applications of liberal education these days are to be found not just in the ivy-covered institutions of the USA but also in the ancient courtyards and quickly rising campuses of Asia. Consideration of new or renovated Asian versions of liberal education affords the chance to see what is most relevant in these methods and what can be adapted for greater success in the future.

One of the first innovations that Yale-NUS introduced was a rigorous common curriculum. This addressed several of the pitfalls discussed above by including texts both Western and non-Western, pairing Confucius with Aristotle, the *Odyssey* with the *Ramayana*, and also bringing modern texts from Asia and the West in conversation with each other. Such a comparative approach is not just limited to the humanities. Courses on "Comparative Social Institutions" and "Historical Immersion" carry this global scope into the social sciences, too. Yale-NUS attempts to bridge the "two cultures" by having all students develop basic scientific literacy. The common curriculum gives a broad and rigorous introduction to the methods of the humanities, social sciences, and the natural sciences. Out of ten required courses, three focus on the natural sciences and one on quantitative reasoning. Rather than offering non-majors watered-down classes, each common curriculum course challenges students to understand a variety of disciplinary approaches. This teaches students to become proficient in and understand the applications of multiple subjects, as well as to bring these together in their work in and beyond the university.

In the course of the past decade, much has been said about the failure of traditional liberal education programs to furnish students with the tools to make wise financial decisions or to vote on hot-topic political issues. By introducing every students not just to "science" or "math," but to the structure of scientific inquiry, Yale-NUS is providing them with the vocabulary and confidence to think critically within their own disciplines and future careers, as well as to become responsible citizens and leaders. At times, innovation means simply possessing the awareness of what might need changing, and a training in all three branches of the liberal arts—science, social science, and humanities—will equip students with precisely this openness.

In building innovation into the core structure of liberal education at Yale-NUS, we have also sought to recruit our faculty as champions of change. By placing faculty in divisions that are inherently interdisciplinary, rather than disciplinary departments, Yale-NUS strives to break down the silos inherited from the traditional nineteenth-century organization of research universities. The result is an integration of disciplines that is emulated in the students' common curriculum. Faculty participate in workshops and teach in teams that help them to generate new ideas regarding both research and pedagogy. Yale-NUS recruited faculty through workshops in which, in addition to presenting their current research, candidates share their conceptions about pedagogy and about extracurricular life. As a result, and with the advantage of self-selection on the part of prospective faculty, Yale-NUS has found an inaugural faculty unusually eager to engage with students both in and outside the classroom.

Yale-NUS has built its faculty and created the common curriculum with one central question in mind: "What must a young person learn in order to lead a responsible life in this century?" In other words, what education must we provide for our students such that they continue to learn and create once they leave our campus? How do we ensure that they do in fact live what Socrates called "the examined life," thinking critically about their own values, and at the same time have the opportunity for an active life, one that allows them to make a difference beyond the campus walls? One solution was already available: by placing the school in a uniquely cosmopolitan city in Asia, Yale-NUS is able to bring together students from a variety of backgrounds, including adventurous students from all over the world. The entering class of 2013, comprising about 150 students, was roughly 40% international, including young women and men from six continents. This sort of diversity allows students to explore new practices and viewpoints simply by working with their peers. It also allows them to embrace risk and the possibility of failure, since students know that they will encounter others with habits, traditions, and languages that are entirely new to them, and that they will have to test and retest methods for finding a middle ground.[7] Because Yale-NUS students have come from far and near, and from a variety of socioeconomic backgrounds, they are primed to become self-reliant and to probe traditional knowledge with open but critical minds.[8] Yale-NUS strives to capitalize upon these instincts through an education that leads them even further beyond their comfort zones. As an international community of learning, the college teaches its students to discover and create new opportunities in their world.

In addition to in-class curriculum, Yale-NUS has designed a program that requires students to bring their campus-bound learning into the world, to make education into innovation. This "Learning Across Boundaries" program developed out of a faculty initiative and relies on close collaboration among faculty, staff, and students. All students at Yale-NUS spend a week immersed in off-campus projects with faculty mentors, on topics such as biodiversity in Thailand, Burmese literature, or Buddhist philosophy in Kyoto, Japan. Much like the "20% time" set aside by major corporations like Google for their employees to pursue self-driven, innovative projects (Tellis 2013, p. 13), these off-campus trips provide students with a canvas on which to experiment with the skills they have learned in class. Students push themselves to understand the world not just intellectually, but practically, to apply their education as a basis for engagement and empathy.[9] By going out into the world as an integral component to the common curriculum, students practice bridging the gap between world and campus, precisely the same bridge that they will cross as they graduate from our institution. Yale-NUS hopes that this practice in bringing liberal education to real-world applications will allow students to replicate and expand upon their experiences in college by creating their own innovations in the world. The college hopes, further, that this will foster a seemingly paradoxical *tradition* of innovation. As Tennyson's Ulysses says (verses 18–21 of "Ulysses"),

> I am a part of all that I have met;
>> Yet all experience is an arch wherethrough
>> Gleams that untraveled world whose margin fades
>> Forever and forever when I move.

Instead of seeing college as an endpoint to their achievements, the hope is that students will understand it as a testing ground from which they can learn to encounter the wider world.

Through a broad but well-defined and intensive common curriculum, the integration of different disciplines, the recruitment of an energetic faculty with a strong commitment to undergraduate education, and finally the drawing together of world and campus, Yale-NUS is striving to create an international community of learning. The community is founded in the conversations facilitated by the liberal education tradition and it addresses head-on the challenges that liberal education confronts as it adapts to the twenty-first century and spreads throughout the globe. Most of all, the community is founded on the idea that we wish to teach students to antici-pate change, to ask future-facing questions, to take on risks, and to carry their learning beyond the walls of the campus. Through interdisciplinary, international knowledge, through self-reliance and teamwork, the college wishes to create a campus of innovators. Yale-NUS has summarized the mission as follows:

A Community of Learning
    Founded by two great universities
    In Asia, for the world.

It is our hope that liberal arts traditions from Asia and the West will con-tinue to enliven Asian educational systems in the generations to come, shaping a generation of Asian leaders who are also citizens of the world.

NOTES

1. The list here is influenced by but not identical with that of Andrew Delbanco in *College*.
2. Frederic Laloux (2014) discusses the efficacy of this non-hierarchical pro-cess in the context of contemporary business (see *Reinventing Organizations*, p. 100–4).
3. On the history of Chinese education (and higher education), see Lee 2000, *Education in Traditional China*.
4. As Laloux explains, this "holocratic" approach is also adaptable to profes-sional contexts (p. 90, p. 119).
5. See Snow (1998) on the polarization of "literary intellectuals" and scien-tists, especially "physical scientists" (p. 4).
6. Indeed, as Michael Roth (2014) recounts in his history of the development of liberal education in the USA, innovation within the liberal arts has often arisen out of the creation of entirely new institutions. Take, e.g., Thomas Jefferson's creation of the University of Virginia, which he chose to pursue rather than striving to reshape the original public institution in Virginia, the College of William and Mary (27).

7. Writers on innovation identify embracing risk and encountering the possibility of failure as key criteria for the creation and implementation of new ideas (see Tellis, 65*ff.*, Black).
8. Roth traces "self-reliance" as a key component of liberal education back to Ralph Waldo Emerson and Booker T. Washington.
9. On Jane Addams' pioneering link between empathy and education, see Roth (2014), p. 18.

REFERENCES

Bowen, W., Tobin, E., & Kurzweil, M. (2005) Equity and excellence in American higher education Charlottesrille: University of Virginia Press.

Confucius. (2014). *The analects* (trans: Chin, A.). New York: Penguin Books.

Delbanco, A. (2012). *College: What it was, is, and should be.* Princeton: Princeton University Press.

Deresiewicz, W. (2014). *Excellent sheep: The miseducation of the American elite and the way to a meaningful life.* New York: Free Press.

Gadamer, H. (2004). *Truth and method.* (trans: Weinsheimer, J. & Marshall, D. G.). London: Bloomsbury. (Original work published 1960).

Laloux, F. (2014). *Reinventing organizations: A guide to creating organizations inspired by the next stage of human consciousness.* Brussels: Nelson Parker.

Lee, T. H. C. (2000). *Education in traditional China: A history.* Boston: Brill.

Lewis, H. R. (2007). *Excellence without a soul: Does liberal education have a future?* New York: Public Affairs.

Pallais, A., & Turner, S. E. (2007). Access to elites. In S. Dickert-Conlin & R. Rubenstein (Eds.), *Economic equality and higher education* (pp. 128–156). New York: Russell Sage.

Roth, M. S. (2014). *Beyond the university: Why liberal education matters.* New Haven: Yale.

Snow, C. P. (1998). In S. Collini (Ed.), *The two cultures.* Cambridge: Cambridge University Press (Original work published 1959).

Tellis, G. J. (2013). *Unrelenting innovation: How to build a culture for market dominance.* Hoboken: Wiley.

Yale University. (2012). *Promoting diversity and equal opportunity at Yale University.* New Haven: Yale University.

CHAPTER 5

New Liberal Arts and Sciences Institutions in India and Singapore: The Role of STEM Education

Bryan Penprase

Abstract India is facing a vast unmet demand for both a greater quantity and quality of higher education. The rise of India's middle class, youth-dominated demographics, and emerging high-tech sectors make the expansion of higher education in India an urgent priority. Private philanthropy has enabled the development of a new sector of private liberal arts and sciences universities. We describe several of these new universities in India, and how they each uniquely embody liberal arts in India. Yale-NUS College as an interesting counterpoint is also described, with its interdisciplinary curriculum that blends East and West. These new universities offer great promise to educate students in ways that are rooted both in the twenty-first century and in the cultures of India and Singapore.

Keywords liberal arts • curriculum development • institution building • STEM Education • Indian education

B. Penprase (✉)
Professor of Science, and Director, Centre for Teaching and Learning, Yale-NUS College, Singapore

Pomona College, Claremont, CA, USA

© The Editor(s) (if applicable) and The Author(s) 2016
W.C. Kirby, M.C. van der Wende (eds.), *Experiences in Liberal Arts and Science Education from America, Europe, and Asia*,
DOI 10.1057/978-1-349-94892-5_5

61

Twenty-First-Century Skills and Indian Higher Education

In global dialogues on liberal arts and science, concepts that define the educated person, often expressed as a list of virtues, are often discussed. The list of virtues may arise from Chinese Confucian or Taoist traditions, or they can be reverse-engineered from successful alumni or entrepreneurs where they often are described as "twenty-first-century skills." These virtues of an educated person form the ultimate learning outcomes, which we should develop carefully in discussion with a wide variety of stakeholders—CEOs, employers, faculty, parents, and funding agencies. These virtues are intended to enable graduates to lead a successful life as it might be defined within the social, cultural, and political environment in which the graduates will live. Accordingly, a dialogue on liberal arts and sciences in the twenty-first century necessarily should include a thorough discussion on the new concepts of higher education being developed in India, home to 1.2 billion people and the site of a burgeoning new sector of higher education: the Indian private liberal arts and sciences university.

The twentieth-century writer F. Scott Fitzgerald said that "the test of a first-rate intelligence is the ability to hold two opposed ideas in mind at the same time and still retain the ability to function" (Fitzgerald 1936). Keats in his piece "On Overcoming Milton" described a form of creativity exemplified by Shakespeare as "negative capability." Negative capability includes a form of intuition that transcends the rational analytic mind by developing *empathy* for its subject, and which tolerates uncertainties "without any irritable reaching after fact and reason" (Pollard 2015). The complexities of life in the twenty-first century demand these virtues more than ever, and such virtues are especially important for higher education within India with its thriving multi-layered, diverse, and complex culture. Rabindranath Tagore argues in his piece on *The Religion of Man* (1930) that progress can only happen through a form of education that cultivates "inclusive sympathy" through "global learning, the arts and Socratic self-criticism"—all elements within modern liberal arts, and all which enable students to understand complexity and diversity (Nussbaum 2011).

India is a place of simultaneous opposites. While India offers some of the most advanced multi-billion-dollar high-tech companies in the world, such as Wipro, Infosys, and Accenture, they co-exist with traditional and rural communities that continue practices from thousands of years ago. India is a country with pockets of unmatched wealth in its cities next door to some of the most desperately poor settlements in the world. Within its road system,

India offers fascinating laboratories of collective behavior that could challenge systems analysts or chaos theorists for decades to come. On the other side of walls that line India's roads are some of the most advanced corporate training compounds, science laboratories and hospitals, as well as some of most serene and beautiful temples, gardens, and homes in the world.

India also has a proud academic tradition that includes some of the first universities in the world, such as the ancient Nalanda University, which housed a vast library and was filled with large numbers of international scholars in Buddhism, medicine, and other subjects from 400 to 1200 CE. India developed several of the first notions of what we might call liberal arts in Vedic times, and is home perhaps to the first Asian liberal arts college, Shantiniketan, founded by Rabindranath Tagore in 1901. Liberal Arts and Sciences in India, much like in the rest of Asia, is of particular interest today, as India faces a wave of growth in its economy (in 2014 growing at 8 % annually) and very young population (in 2014 growing at 4.5% annually) (British Research Council [BRC], 2014). India's rapid economic and demographic growth has sparked many urgent environmental issues, as well as a massive unmet need for higher education to develop a well-trained workforce with what we have been calling "twenty-first-century skills" (BRC 2014).

A few numbers can help further sharpen the dilemmas faced by India's higher education system, which has interesting comparisons with China. India, like China, has a very large population. India's population currently stands at 1.22 billion (compared to China's 1.37 billion), and is growing rapidly to overtake China in the coming decade. India already has the largest population of under-20 individuals in the world, approaching 300 million by the year 2030 (Federation of Indian Chambers of Commerce and Industry [FICCI], 2013). India, like China, has a rapidly expanding middle class, which by some estimates exceeds 300 million. Like China, India has an extremely strong cultural emphasis on exam-based proficiency and academic success, with "toppers" (those who make the best national exam scores) making headlines within newspapers and becoming minor celebrities in their hometowns. India's higher education system has attempted to expand very rapidly and strains to meet the demands of the new middle class, as it aspires to educate their children. Like China, India has about 17–20% of its young people enrolled in some form of tertiary education, and has a need to rapidly expand its higher education system—perhaps by more than double—to fully meet this demand (American Council on Education and Center for International Higher Education 2013).

Unlike China, India lacks efficient and centralized strategic planning. India faces massive shortages of teachers and professors, with shortages of secondary and tertiary professors exceeding one hundred thousand and large drop-out rates that approach 30% by grade 8 (BRC 2014). Together, these factors prevent delivery of higher education or even a high school education to a vast majority of India's large population. Amartya Sen, describes in his book, *The Argumentative Indian*, some of the complexities of achieving consensus within India, in which "heterodoxy is a natural state of affairs" and how argument and prolixity are intrinsic aspects of India's public life for over a thousand years (Sen 2005). This means that with 1.3 billion Indians there are nearly as many opinions within India about how to reform and expand its education system. All however would agree on the need for resources—which are vastly less than what is needed for developing rural schools, for training teachers, and for creating the types of excellent holistic education and active learning on a wider basis.

The recent election of Narendra Modi, widely applauded in the USA Indian Non-Resident Indian (NRI) community, has yet to yield dividends for India's higher education system. Modi has been labeled as a reformer who can energize and spark the kinds of changes needed within India's education system. Modi's government has discussed abolishing the widely criticized Higher Education Planning Commission, also known as the University Grants Commission or UGC, which regulates higher education. But as yet there has been little progress in the promised increases in autonomy and resources to universities (Kohli 2015). Within the two different bodies that govern Indian higher education, the UGC and the National Knowledge commission, there are multiple opinions and sometimes contradictory recommendations that impede reform or the founding of new institutions (Tharoor 2015). The much needed expansion of India's university system, which could include five new Indian Institute of Technology (IIT) campuses, has yet to be implemented, and also seems to be underfunded by the government. According to one source, education funding had increased by 18 and 8% in the previous two years, but in the current budget faces a lower percentage of growth than India's GDP expansion (Narayanan 2015).

India is well known for its excellent IIT graduates, who in many cases (like MIT graduates) form companies that have fueled much of India's economic growth. IIT graduates emerge from a nearly life-long selection process and represent the top 0.1 percentile of India's student population, with less than 2% of IIT applicants being selected. The intense selectivity is partly due to the low capacity of the IIT system, which in its 16 campuses are able to educate only 10,000 undergraduate students each year, after receiving applications from about 500,000 students.

To attempt to meet the demands for increased capacity of higher education, India has built the largest number of educational institutions in the world. These institutions all face a very complex regulatory environment governed by the UGC and span a vast range in quality. In 1956, when the UGC was created, there were about 30 universities within India. This number grew to 200 in 1990, and is now at approximately 666, with over 14,000 colleges, which in the Indian context refers to institutions that grant degrees in grade 12 (BRC 2014). The expansion in number of institutions has included the development of a large number of for-profit, private institutions that are of very low quality, and even the best of India's universities lag behind in international rankings. An oft-cited statistic within India is that none of the top 200 world-ranked universities are Indian, and among the top 300 only two Indian institutions are included—one IIT campus and the University of Delhi (Hindustani Times 2013).

One solution to creating high-quality higher education in India is the growth of new private, liberal arts, and sciences universities, founded through partnerships between philanthropists and high-tech entrepreneurs. Indian technology leaders often bring hybrid international perspectives to their approach toward higher education, as many received graduate degrees abroad after growing up within the Indian secondary education system, and in some cases the IIT undergraduate system. New institutions are being founded to address some of the unmet needs of the Indian companies, who are demanding agile and innovative employees. These institutions also promise to help meet the challenges of Indian society, which requires graduates with a broad and deep knowledge of India's cultural history, politics, and arts as well as deep proficiency and facility with STEM subjects. Since most of these new institutions are being funded by private money, they are able to implement new forms of curriculum and faculty governance without government regulation and interference. At least eight of these new universities have been founded since 2011, and they provide a diverse and exciting range of new models for liberal arts and sciences in India.

To help foster a dialogue among these emerging institutions within India, and to facilitate collaborations internationally with North America and Asia, the author co-organized two conferences within India on the "Future of Liberal Arts and Sciences in India." The first of the meetings was in Bangalore during January 2014, and was hosted by the Raman Research Institute (RRI) and Indian Institute of Human Settlements (IIHS). The meeting featured the founding Vice Chancellors and Presidents of eight new Asian liberal arts institutions, as well as Presidents and Chancellors from two University of California (UC) campuses and some of the best liberal arts colleges in the USA. The meeting also featured leaders within India from government, large corporations, non-governmental organizations (NGOs), and even new private high schools within India (Penprase 2015a). The first meeting discussed the array of issues facing higher education in India, and also set the stage for a second meeting in March 2015 in New Delhi. The second meeting convened on the new campuses of two new Indian liberal arts institutions described below: Ashoka University and the O.P. Jindal Global University. The group additionally convened for a day at the new University of Chicago center in New Delhi. Materials from both conferences will be added to a website for an online conference proceedings that should be a valuable resource for understanding Indian liberal arts and sciences within India (Penprase 2015b). Below is a brief overview of some of the properties of a cross section of these remarkable new higher education institutions within India, with short synopses to give a sense of the institutions, their founders, and their different approaches to higher education and liberal arts in the Indian context.

Ashoka University was founded by a group of US-educated entrepreneurs, such as the Yale alumnus and philanthropist Ashish Dawan, who also created the Central Square Foundation dedicated to providing education to children within slums in India. Ashoka University aspires to be the "Yale of India." Among the founders of Ashoka University is Pramath Sinha, founding Vice Chancellor of Ashoka. Pramath was educated at an IIT, and previously founded the Indian School of Business, as well as a business partnership with Madeline Albright. The Ashoka University founders began by establishing the "Young India Fellowship," a one-year multidisciplinary postgraduate program in Liberal Studies and Leadership (Ashoka University 2015). The Young India Fellowship includes a year of study with lectures from a wide variety of disciplines in humanities, sciences, and arts. Fellows complete a project at the end of the year that blends liberal arts with an internship and mentoring from the instructors in the program. The program has grown in its three years to 200 fellows from its initial batch of 58 students, and is now housed at the new Ashoka University campus. The Ashoka University opened in 2014–15 with its first class of 65 men and 68 women, and it is admitting its second batch this year in 2015 (YIF 2015). The Ashoka curriculum includes 12 Foundation Courses, which offers a diverse mix of sciences and humanities. Students then choose a major, which consists of 12–16 courses in about 12 different fields. Ashoka University has developed its curriculum with academic partnerships with

Carleton College, Sciences Po (France), Penn Engineering, the University of Michigan, and King's College (UK). The Ashoka STEM curriculum includes Principles of Science (focusing on ways of knowing and scientific inquiry), Mind and Behavior, and Introduction to Mathematical Thinking. The plan is for Ashoka to grow beyond 2000 students in the coming years, with both undergraduate and graduate programs.

The O.P. Jindal Global University was founded in 2011 and has five graduate programs, with a newly created undergraduate program in Liberal Studies and Humanities. Raj Kumar, its founding Vice Chancellor, returned to India after an academic odyssey that included a Rhodes scholarship at Oxford, a L.L.M. degree from Harvard Law School, and teaching jobs at the University of Hong Kong and the University of New Delhi. Raj's experiences with international higher education inspired him to establish an institution in India with a similar focus on excellence. Steel magnate Naveen Jindal contributed funds for the expansive campus, and named the institution after his father, O.P. Jindal. The school is best known for its excellent law school with strong programs in international justice and human rights, and also has robust graduate programs in business, international affairs, as well as government and public policy. A dual degree program has been developed with Rollins College in Florida. Jindal undergraduates will complete two years at Jindal, and then receive a dual degree from Rollins—enabling Indian students to receive a liberal arts and sciences education in both countries. Other academic partners for Jindal include Indiana University-Bloomington, University of Arizona, University of California-Berkeley, Tilburg University (for joint law degrees), and the University of Texas at Dallas for a joint Business degree.

Shiv Nadar University (SNU) was founded in 2011 with the help of the eponymous Shiv Nadar, who made his fortune after founding the tech company HCL. The Founding Vice Chancellor, Nikhil Sinha, assembled in just four years a complete university with graduate and undergraduate programs on a 286-acre campus. SNU is one of the older and larger of these new Indian institutions, with a total of 1250 undergraduates, and 250 graduate students. The institution conducts significant science and engineering research, and offers PhD programs in several of the natural sciences. The schools of SNU include Engineering, Humanities and Social Sciences, Law, Management and Entrepreneurship, and Natural Sciences. Academic partners include Duke University, University of Pennsylvania, and Babson College. One of the innovative partnerships at SNU is a 2+2 partnership with Carnegie Mellon University (CMU). In this program, students will take their first and third years at SNU, and then their second and fourth years at CMU. At the end, students would then be awarded two degrees (one each from both SNU and CMU) with full ABET accreditation. A further innovation for SNU is that it offers merit-based, need-blind admissions, using high school grades but also a special aptitude test, like the SAT but custom-made for SNU by Pearson Education. The admissions office also includes essays and interviews of all of the candidates as part of the process. The faculty are involved in this interviewing process, making them fully part of admissions, instead of having it outsourced to a separate department. SNU has already become a selective institution in its short history—with thousands of students applying for admission, and less than 25% being selected. It will graduate its first batch of undergraduate students in 2015.

IIT Gandhinagar (IIT Gh) is one of the newest of the famed IIT campuses, founded in 2008 in Ahmedabad in Gujarat province. While not a true liberal arts institution, IIT Gh has some novel features for an Indian university, as it blends humanities and arts into the studies of all of its young engineers. IIT Gh offers four-year BTech and MTech programs in five branches of engineering as well as two-year MSc programs in several science areas. Ph.D. programs are offered in seven areas of engineering, as well as in the four physical sciences. IIT Gh also has a 2-year MA program in Society and Culture, and offers PhD degrees in History, Literature, Philosophy, Psychology, Sociology, and Social Epidemiology. The founding director, Sudhir Jain, has intentionally brought a mix of liberal arts, critical thinking, and humanities into his IIT campus. One of the most innovative features within IIT Gh is a five-week Foundation Program, which is offered to a mix of overseas students from the California Institute of Technology and the entering class of Indian IIT students. The program is known as *India Ki Khoj or "Discovery of India"* and has "been designed to help overseas students relate with this idea and the many layers that form identity in India" (IIT Gh 2015). The program works with a combination of academic topics and field trips to "communicate the many imaginations about India" and to "take students through the India of the past, present and future" and to connect India's ancient traditions of philosophy to its present (IIT Gh 2015).

Azim Premji University is another example of private philanthropy creating a new form of higher education in India. The Azim Premji University is the most recent project for the multi-billion-dollar Azim Premji Foundation (APF), which is dedicated to creating a "just, equitable, humane and sustainable society" within India by "making deep, large scale and institutionalized impact on the quality and equity of education in India, along with related development areas" (APF 2015). Azim Premji, the namesake of both the University and the Foundation, was educated with a B.S. in Electrical Engineering from Stanford University, and began his business career in 1967 at the young age of 22, when he took over the Wipro Limited Corporation. Wipro is now a multi-billion-dollar technology company with 2013 revenues of $6.9 billion and over 145,000 employees serving clients in 57 countries. Azim Premji is one of the world's wealthiest men, listed as 41st richest in the world, with a personal fortune of $17.2 billion. He founded APF in 2001, with the goal to contribute to high-quality and universal education in India. The foundation funds pilot programs in public schools in India, and pledged to contribute $2 billion to have a major impact on the 1.4 million schools within India. Azim Premji also has joined Bill Gates "Giving Pledge" program to give most of his wealth to charitable causes along with Warren Buffet and Richard Branson (Wipro 2015). Azim Premji University programs are designed to create talent, knowledge, and also social change, and are focused in several interdisciplinary programs. Azim Premji University offers Master of Arts in Education or Development, conducts research in education and development, and houses a continuing education center for teachers. The Azim Premji curriculum includes interdisciplinary explorations of socially relevant themes. Examples include courses in "Law, Governance and Development," "Mind and Society," "People and Ideas," and "The Philosophy of Education." The undergraduate program at Azim Premji is similarly interdisciplinary and will open with its first batch of students in 2015. It will offer major concentrations to undergraduates in four areas—Physics, Biology, Economics, and Combined Humanities—and in each of the majors, the curriculum emphasizes the connections between these disciplines and India's culture and society (APU 2015).

The IIHS in Bangalore aims to both study India's rapid growth and shape this growth by training experts in "Urban Practice" through a master's degree program. IIHS hopes to "address challenges of urbanization through an integrated program of education, research, consulting and advisory services." IIHS also plans to expand into a full-fledged university, perhaps with multiple campuses, to provide both undergraduate programs and PhD degrees. The IIHS University expects to build schools of environment and sustainability, human development, economic development, governance and policy, and settlements and infrastructure. At present, IIHS offers the Master's Degree in Urban Practice, and is awaiting recognition from the Indian government to initiate a bachelor's undergraduate program in the same area. IIHS has been funded by grants from Bangalore's high-tech entrepreneurial community, with multi-million dollar grants from Nandan Nilekani, former CEO of Infosys. The founding director of IIHS, Aromar Revi, describes how the IIHS can help understand India's complexity, especially in its rapidly expanding cities. According to Aromar, "the most complex systems that have been built by human beings are cities. We have lived and worked in them for 5000 years, but we don't know how to manage them at the scale necessary in a largely urban world" (Revi, personal communication, 2013). The IIHS strategy is on two levels: to train practitioners from companies, government, and civil society organizations through short courses, and to develop a "cloud of practitioners" who can help guide projects that both educate students and help solve local problems via their practice. IIHS also hopes to train graduates who are able to answer not only technical questions (the "what, how and when") but find the deeper causes of problems by asking "why" systems are in their current state. The IIHS provides a laboratory for studying and helping solve India's problems, which can then be used to help other countries such as Brazil, Mexico, and China that have similar issues of sustainability and urbanization. According to Aromar Revi, "India has a monopoly on some of the most complex problems that the world may face in the 21st century" and the IIHS is working to develop the necessary theory, perspectives, and practices to help address these problems in a way that can possibly be adopted for other countries in the coming decades (Revi 2013).

The final institution this essay will examine is **Yale-National University Singapore (NUS) College in Singapore**, which, while not in India, provides another new model for higher education in South and Southeast Asia. Yale-NUS College has been actively involved in both of the "Future of Liberal Arts in India" conferences and is beginning to collaborate with many of these newly developed Indian institutions. The curriculum at Yale-NUS College was designed to answer the question: "What must a young person learn in order to lead a responsible life in this century?" The curriculum was designed after a careful study of the history of higher education, such as the Yale curriculum report of 1828, with its metaphor of building the "discipline and furniture of the mind." The Yale-NUS College offers a broad and deep common curriculum that blends Eastern and Western works and that integrates science and quantitative work in the study of all students. Within the courses is "a focus on articulate communication," and "open, informed and reflective discourse" (Garsten et al. 2013). Students at Yale-NUS College take nearly two years of a common curriculum to provide a common base of understanding and discussion. These courses include a full two years of STEM subjects, beginning with Scientific Inquiry, which explores the ways in which science and math conceptualizes and discovers aspects of nature; Quantitative Reasoning, which combines social sciences, programming, and statistics; and a two-semester course entitled Foundations of Science, which examines a range of scientific topics from disciplinary perspectives and then combines students into common experiences that include field work in Malaysia and developing answer to "Grand Challenge" questions on global warming and disaster response (Penprase 2015c). The Yale-NUS College science curriculum is designed to provide an interdisciplinary education that will enable students to help solve complex problems facing Asia and the world in 2015, and that will prepare its graduates to be responsive and informed citizens and policy makers.

From the above discussion, it is apparent how a diverse range of new institutions being developed in India and Singapore are redefining liberal arts and science education in South and Southeast Asia. These new institutions are not adopting an American model but reinventing a form of education that has been in practice in Asia for centuries—in India, in China and elsewhere. As such, these new institutions in Singapore and India are building on their campuses a form of education that Tagore envisioned in his essay entitled "An Eastern University" (Tagore 1922):

> For our universities we must claim, not labeled packages of truth and authorized agents to distribute them, but truth in its living association with her lovers and seekers and discoverers. Also we must know that the concentration of the mind-forces scattered throughout the country is the most important mission of a University, which, like the nucleus of a living cell, should be the center of the intellectual life of the people.
>
> Rabindranath Tagore, "An Eastern University"

REFERENCES

American Council on Education & Center for International Higher Education—International Briefs for Higher Education Leaders. (2013). India—The next frontier. Retrieved from http://www.acenet.edu/news-room/Documents/International-Briefs-2013-April-India.pdf

Ashoka University. (2015). Founders. Retrieved from http://www.ashoka.edu. in/About/Founders

Azim Premji Foundation. (2015). About us. http://www.azimpremjifoundation. org/About_Us

Azim Premji University. (2015). Academic vision. Retrieved from http://azim-premjiuniversity.edu.in/SitePages/academic-vision.aspx

British Research Council. (2014). Understanding India: The future of higher education and opportunities for international cooperation. Retrieved from http://www.britishcouncil.org/sites/britishcouncil.uk2/files/understanding_india_report.pdf

Federation of Indian Chambers of Commerce and Industry. (2013). Vision 2030. Retrieved from http://www.ey.com/Publication/vwLUAssets/Higher-education-in-India-Vision-2030/$FILE/EY-Higher-education-in-India-Vision-2030.pdf

Fitzgerald, F. S. (1936, March). "Pasting it together", part two of The Crack-Up. In *Esquire*.

Garsten, B., Patke, R., Bailyn, C., Jacobs, J. J., Chuan, K. H., & Penprase, B. (2013). Yale-NUS College—A new community of learning. Retrieved from http://www.yale-nus.edu.sg/wp-content/uploads/2013/09/Yale-NUS-College-Curriculum-Report.pdf

Hindustani Times. (2013). A vision for education—From 2013 to 2030. *Hindustani Times*. Retrieved from http://www.hindustantimes.com/chunk-ht-ui-hteducationsectionpage-otherstories/a-vision-for-education-from-2013-to-2030/article1-1153460.aspx

Indian Institute of Technology, Gandhinagar. (2015). India Ki Khoj. Retrieved from http://www.iitgn.ac.in/india-ki-khoj/.

Kohli, G. (2015, April 16). Private universities ready for choice-based credit system. *Hindustani Times*. Retrieved from http://www.hindustantimes.com/higherstudies/private-universities-ready-for-cbcs/article1-1337474.aspx

Narayanan, N. (2015). Whether the UGC is scrapped or revamped, it has failed India's higher education. *Scroll.in*. Retrieved from http://scroll.in/article/718511/whether-the-ugc-is-scrapped-or-revamped-it-has-failed-in-its-mission

Nussbaum, M. C. (2011). Democracy, education and the liberal arts: Two Asian models. *University of California at Davis Law Review, 44*, 735. Retrieved from http://chicagounbound.uchicago.edu/journal_articles/3302/

Penprase, B. (2015a). Liberal arts in India [blog post]. Retrieved from http://bryanpenprase.org/liberal-arts-in-india/

Penprase, B. (2015b). The future of liberal arts in India 2015 [conference website]. Retrieved from http://future-liberal-arts-sciences-india.commons.yale-nus.edu.sg

Penprase, B. (2015c). Foundations of Science at Yale-NUS overview [webpage]. Retrieved from http://fos1aug2015.courses.yale-nus.edu.sg

Pollard, D. (2015). Keats: On overcoming Milton [blog post]. Retrieved from http://www.davidpollard.net/keats-overcoming-milton/
Revi, A. (2013). Personal interview for Penprase, B. *The expanding universe of higher education* (in preparation), chapter available upon request.
Sen, A. (2005). *The argumentative Indian*. New York: Penguin.
Tagore, R. (1922). *Creative unity*. New York: The Macmillan Company.
Tharoor, S. (2015). How Modi government is undermining Indian education. *NDTV.com*.Retrieved from http://www.ndtv.com/opinion/how-modi-government-is-undermining-indian-education-759854
Young India Fellowship. (2015). About YIF [webpage]. Retrieved from http://www.youngindiafellowship.com
Wipro. (2015). About Wipro [webpage]. Retrieved from http://www.wipro.com/about-wipro/

Polymathy, New Generalism, and the Future of Work: A Little Theory and Some Practice from UCL's Arts and Sciences Degree

Carl Gombrich

Abstract It is a truism that we are at the beginning of a revolution, one that is driven principally by technology but also involves other factors such as globalization and problems of planetary scope. Graduate work, too, is changing. More nations are becoming knowledge economies in which services dominate and attributes such as creativity, flexibility, and collegiality are valued in white-collar and professional jobs at least as much as academic subject knowledge.

This chapter sketches a trajectory of higher education in its relation to employment and argues that we see a re-emergence of polymathy and generalism as both valued educational ambitions and central to the future of work. Examples of University College London Arts and Sciences student profiles are given and experiences of graduate recruitment examined.

Keywords Polymathy • generalism • knowledge economy • future of work • expertise • specialization • liberal arts and sciences

C. Gombrich (✉)
University College London, London, UK

© The Editor(s) (if applicable) and The Author(s) 2016
W.C. Kirby, M.C. van der Wende (eds.), *Experiences in Liberal Arts and Science Education from America, Europe, and Asia*,
DOI 10.1057/978-1-349-94892-5_6

*Each individual becomes a unique personality by synthesizing the dispa-
rate things s/he learns. This is the perennial value of liberal education.
What makes a being a 'person' is that they're a whole that's more than the
sum of their disciplinary skill-sets. Given that in our volatile Neo-liberal
environment no single skill-set is any longer a guarantor of a lifelong
career, this training of the person becomes more essential than ever before*

Steve Fuller

INTRODUCTION

The main claim of this chapter is that a liberal arts and sciences education
of some stripe is the best sort of education for a large and growing proportion of white-collar jobs across the globe. It surely must be appropriate to
investigate what our students do after graduation—indeed, it would be
negligent not to—even if some of us in universities express doubts about
serving 'business' or pandering to 'neo-liberalism'.

Here, liberal arts and sciences will mean that sort of education usually
taken to have begun with the writings of Varro and Cicero in Ancient Rome
and later developed into the seven liberal arts by great medievalists such at
Martianus Capella and Boethius. Such an education incorporated what we
would now call both humanities and sciences. This concept of education
dominated Western universities from the founding of Bologna up until the
rise of specialist technical institutions in France in the nineteenth century,
and was recast first by Humboldt at the University in Berlin in 1810, and
then in the USA where it survives today as liberal arts in many elite institutions. Such a conception is, of course, immensely broad and has undergone
very significant evolution, but at its heart are ideas of holism (even uni-
versalism) and interconnection between parts of knowledge (The BPTI's
Channel 2012). This distinguishes it from education in single academic
disciplines or solely technical subjects which has become the norm for the
large majority of students in the UK and a great many students throughout
Europe and in China, India, and the rest of the developing world.

The usual defence of the liberal arts in modern times takes one or more
of three tacks: (1) the value of the humanities versus science; (2) the inherent value of education versus the instrumental[1]; and (in a related vein) (3)
the value of educating people as self-realized, critical citizens as opposed to
producing economic agents for society—an entity which, to the dismay of
many instinctively opposed to seeing the world as a market, some now call
simply an 'economy' (Deresiewicz 2014; Michaels 2011; Nussbaum 2010).

As we have noted above, in the wider tradition the arts and sciences were never separated and, indeed, it is a mistake to do so. Therefore, the first defence of the liberal arts will not be addressed here.

Regarding the second point, the supposed clash between inherent and instrumental values, I have written elsewhere on how I see a melting away of this dichotomy (Gombrich 2015a). This is positive for arts and sciences education. Many things of inherent value which have traditionally been associated with such an education—creativity, empathy, critical thinking, love of learning, collegiality—are now precisely the things that business leaders are telling us they look for in graduates (Hagel 2009). This is an important shift in the meaning of 'instrumentality' as it problematizes the usual distinction between education for life and education for work. Here, however, I will focus on a related but slightly different theme: the relationship between the subject matter of higher education and getting a job afterwards.

THE EMPLOYMENT OF UNIVERSITY GRADUATES

What sort of work will our graduates do?[2] It is widely acknowledged that in the next 10–20 years the G20 countries will become 'knowledge economies' (Chen and Dahlman 2005; Deloitte 2014). Such economies do not rely principally on mining, making tangible products, or growing materials. Instead, they rely on 'intangibles' such as intellectual property, shares, or digital or virtual goods (Jarche 2013). Most definitions state that knowledge economies involve the interaction of people with IT (Powell and Snellman 2004). The 'knowledge economy' is still, perhaps, a problematic concept (Brown et al. 2011; Robertson 2003), but few seriously deny that collectively the sorts of jobs our children will do in these economies will be very different from the jobs of our grandparents—or even our parents. Robots and Artificial Intelligence are poised to encroach enormously—obviously in manual labour and logistics but also in many of the professions: in law, medicine, and accountancy (Economist 2014; Susskind 2013). More positively, the explosion of technology, the ubiquity of the internet, the increasing interaction between cultures and the increased mobility of people will throw up countless new jobs which we cannot currently describe well.

Lacking the fine grain, one way we classify such jobs is as 'services'. In the UK, the Office of National Statistics estimates that already about 80 % of workers are in services (Office of National Statistics 2013). In China, too, services are growing fast, both absolutely and as a share of the economy (New York Times 2015). Jobs in services involve sophisticated interactions with other people and may be more or less technical.

Here are some examples: 'Platform Interface Advisor for Clothing Retail Company'; 'Local Government Digital Health Engagement Officer'; 'Head of Transparency for National Travel Company'. These jobs are easy to parody, much as anything new can appear faddy or ridiculous. In England, there are two excellent TV series lampooning such roles,[3] and the anthropologist David Graeber makes a more serious critique with his idea of 'bullshit jobs' (Graeber 2013). But earlier generations can rarely imagine or see the point of what their successors might do,[4] and there is no reason to think that this time it is any different.

One thing that is striking is the tenuousness of the connection between these sorts of white-collar, middle-class jobs and traditional academic disciplines. It is not that they are not highly skilled—they are—but they do not correspond well to 'History', 'Literature' or even 'Biology', or 'Computer Science'. This has led to a situation in which, conservatively, 50% of graduate jobs do not demand any particular disciplinary qualification (Open University 2015). Although this statistic is from the UK and depends somewhat on the historic flexibility of that labour market, it is likely to rise in all industrialized countries when the connection between academic disciplines and white-collar contemporary work is recognized to be weak and getting weaker.

Why this gap between academic specialization and the current world of work? The answer, I think, is both simple and profound: we are entering an age in which greater connectivity (of information, products, and people) and increased complexity bring significant and frequent changes to what is required in work outside universities. However, academic disciplines, established in the West roughly during the late eighteenth and nineteenth centuries and now deeply ensconced in university cultures and organizational structures, can find it hard to mirror these changes and to respond.

Many have written on the historic establishing of academic disciplines (see e.g. Schaffer 2013; Shumway and Messer-Davidow 1991) and others have remarked on the connection between this and the birth of industrialized, professionalized economies (e.g. Haskell 2000; Jarausch 2004). Broadly speaking, the rise of science and the professionalizing of academic work in Europe and the USA gave rise to specialist journals and a gradual 'siloization' of academia; meanwhile, and in parallel, an economy of industrial specialization, grounded in Adam Smith's division of labour and built on the new factories, rose and reached its apogee in Henry Ford's conveyor belt of mass production and managerial Taylorism in the twentieth century. To use an anti-postmodern phrase, a writing of the history of this period might claim that the 'tenor of the age' was one of specialization, siloization, and certain forms of professionalization influencing all walks of life.

Now contrast our twenty-first-century knowledge economies: there are relatively few people working on conveyor belts; and where do disciplinary boundaries begin and end when two clicks takes you from history to statistics to computer science, or from art to biology to ethics? In academia, there is talk of open access, citizen science and 'new amateurism' (Gombrich 2013) all of which challenge disciplinary boundaries. In the world of work, older professions, as indicated above, are under pressure. Careers are multiple and complex (Sanders and Sloly 2012). Note that computers are expert specialists; they thrive on the division of labour. At a fraction of the cost, they can do work broken down into finite tasks and can master an increasing number of the specializations of old. These specializations are therefore losing both economic and social value. Instead, very human traits such as analogizing, creativity, dot-connecting, and empathizing are at a premium. The best-selling author and serial entrepreneur Tim Ferris, capturing the demise of specialization as a necessity for a successful career, quotes the author Richard Heinlein as saying, 'specialization is for insects'. Ferris continues:

> Is the CEO a better accountant than the CFO or CPA? Was Steve Jobs a better programmer than top coders at Apple? No, but he had a broad range of skills and saw the unseen interconnectedness. As technology becomes a commodity with the democratization of information, it's the big-picture generalists who will predict, innovate, and rise to power fastest. (Ferris 2007)

What we are witnessing in this new world of work is, in Harold Perkins' phraseology, a 'fourth revolution'[5]; one more 'turn of the [historical] screw' (Perkins 1996, p. xii) which took us from lives rooted in agriculture, to industrialism, and then to office work, and to what Perkins calls 'The Third Social Revolution', the era of 'professionalism'.

Perkins' analysis is interesting because, although published only in 1996, it is perhaps best viewed as a description of the end of an era—the era of a certain type of professional elite—rather than charting the continued rise of one. Well-substantiated as his analysis is, Perkins can track only the rise of professional work and services throughout the twentieth century and perhaps did not foresee the flourishing of possibilities in the knowledge economy that the internet has brought. He conflates 'specialism', 'professionalism', and 'expertise' in ways which look problematic 20 years later. For example, we can tease apart the notions of 'expert' and 'professional'. The relationship between these two categories is complex. One can certainly be an expert without being professional (I may be an expert amateur lepidopterist or an expert in the history of Arsenal Football Club) and many jobs which we might now classify as professional (such as those given as examples of services jobs above) are not expert in the sense of requiring long training—rather they require one to become expert in them while performing them over time in the workplace. 'Expertise' is a category under pressure from the ubiquity and abundance of knowledge. Expertise is, of course (!), a good thing, but descriptions of expertise associated only with academia or professional bodies are too narrow to capture most of the requirements of contemporary white-collar work (Gombrich 2015b).

This debate about expertise had started by the end of the twentieth century. In 1994, two years before Perkins wrote, 'the modern world is the world of the professional expert' (1996, p. 1) other scholars had written of 'The Death of the Expert' in their (professional) field of Operational Research (White and Taket 1994). Theirs is an avowedly postmodern take, but it was prescient. The theme of the end of traditional expertise is taken up more recently by the internet historian David Weinberger. Weinberger lists six ways in which expertise is changing: from masterable domains to messily connected topics; from certainty of conclusions to open enquiry; from opacity to transparency; from 'one-way' expertise to 'multi-way interactivity'; and from the individual to the network (Weinberger 2011).

Weinberger comments that this transition to new ideas of expertise is 'messy' and perhaps this transition was difficult to perceive even 20 years ago, at the time of Perkins' book. But given the shifts in our understanding of the concepts of 'professional' and 'expertise', the term 'services' is perhaps more useful than either to describe work trends in knowledge economies. Perkins is certainly aware of the importance of the rise of services as a category of work, and documents this rigorously. Further, he is aware of a 'catch-all' quality of the category 'services' and writes of the 'scarce resource of expertise in its *manifold forms* [emphasis added]' (p. 1) and that 'professional knowledge is based on…education and experience *on the job* [emphasis added]' (p. 6); but it would be too much to expect him to appreciate just how varied service jobs were soon to become. As noted above, previous generations can rarely conceive of what the next one will do: the title 'Head of Transparency' would, indeed, be hard to comprehend before the advent of social media.

However, labour in knowledge economies is now embarking on a new phase of distributed and networked organizational structures, flexible working practices, portfolio careers, networked expertise, and the possibility to re-train in different areas several times in a lifetime. All such developments put pressure on any specialization or expertise which does not permit sufficient flexibility in thought and working practices. It would be naïve to argue against one of Perkins' principal claims that elites of corporate managers, financiers, and other categories of white-collar workers now wield great transnational power and amass great wealth; further we are likely to see pushback from several professional bodies, concerned to protect vested interests and, in many cases, to preserve existing standards. In particular, we may see a pushback from the 'classic' regulated professions such as Medicine, Engineering, and Law, although even these may soon come under pressure (Susskind and Susskind 2015). Knowledge revolutions tend to remake institutions in considerable and unforeseen ways, and we can expect dramatic restructuring and a questioning of any expertise which is unable to cope with complex new demands.

In a 'global dialogue', we must not overlook strong regional differences in work trends. In Germany, for example, there are tighter disciplinary and work boundaries—and Germany does very well in the global economy. Further, knowledge economies may take different forms in places such as India and China. Nevertheless, the events described here will impact all economies, as the current knowledge revolution is of global scope. It is unlikely that the work of the graduate managerial class of Deloitte, Goldman Sachs, or Pfizer will be that different in Shanghai, London, or Delhi. Indeed, corporate leaders in China and India have stated explicitly the need for 'global graduates', able to work successfully across multiple cultural boundaries and work sectors.

EDUCATION, THE FUTURE OF WORK, POLYMATHY, AND GENERALISM

The implications for education of this changing work environment are of course enormous. We cannot determine precisely what this new world will look like. In one sense, all we can do is hedge. But we can hedge better if we start with the realization that the old specialisms will not cut it for a large and growing proportion of our students.

The way we have chosen to hedge at University College London (UCL) on the Arts and Sciences BASc programme[6] is to insist that all students take some combination of social sciences or humanities and sciences or mathematics, throughout their study programme. Alongside a suite of inter-, trans-, and post-disciplinary Core modules, students must choose major and minor combinations which straddle both sciences and humanities or social sciences.

Consider again the job titles[7] above. All would benefit from various levels of digital literacy and computing knowledge, one of them would benefit from health sciences, and all would benefit from ability in writing, speaking, critical thinking, and 'softer' skills associated with humanities or social sciences. But these are just three examples. We find this combination of science and non-science in many areas: for example, the fast-growing area of manu-services[8] (Work Foundation 2010); in most parts of government; and in jobs in inherently interdisciplinary fields such as sustainability, health, and transport.

Perhaps it is a little grand to call an education that spans the non-sciences and the sciences one in 'polymathy', in the way indicated in the title of this chapter. The word carries a certain weight and intellectual expectation. However, Wikipedia defines a polymath simply as 'a person whose expertise spans a significant number of different subject areas' and continues, 'such a person is known to draw on complex bodies of knowledge to solve specific problems'. This is surely in line with our desire to educate sophisticated workers for a complex knowledge economy. Other definitions of polymath—for example at Merriam Webster: 'Someone who knows a lot about many different things'; or the online OED: 'A person of great or varied learning; a person acquainted with many fields of study; an accomplished scholar', define a higher education which, all else being equal, we would wish for our students. Polymathy was natural for the great intellectuals, inventors and industrialists of the fifteenth to nineteenth centuries (Burke 2010). We are all familiar with the iconic figures of Da Vinci and Galileo but one can also cite Thomas Young, Alexander von Humboldt, Denis Diderot, and indeed countless others. Championing polymathy thus allows us to reclaim some classic educational values while educating in skills and knowledge which are desirable in a modern knowledge economy.

'Generalism', is a term which fell out of favour in the twentieth century and towards the end of the long period of specialization described above, but it too has a respectable history in academia and outside. Isaac Barrow, Newton's predecessor at Cambridge, said, 'He can hardly be a good scholar who is not a general one' (Barrow et al. 1700, p. 220), and more recently the writer and critic Lewis Mumford 'used to describe himself as a "generalist" and surely merited that title, since he "bridged the seemingly disparate disciplines of architecture and planning, technology, literary criticism, biography, sociology and philosophy"' (Burke 2010, p. 16). The award-winning writer, explorer, and poet Robert Twigger—who has interesting things to say about polymathy, both as an attitude for a successful working life, and as the basis for an educational programme (Twigger 2015)—makes a distinction between polymathy and generalism. He is positive about the former but somewhat disparaging of the latter: '[A generalist] is usually just a manager of specialists, half in the dark himself, good with jargon and that's about it'(Twigger 2011). However, there are other contemporary writers, especially from the business community, who make a more positive case for generalism. Richard Martin, the consultant and author of a forthcoming book on 'neo-generalism', speaks of generalists as 'people who have the potential, the attitude and the aptitude to specialize in more than one discipline' and notes, in a phrase that has relevance to portfolio careers in a knowledge economy, 'For the generalist, there is something hugely appealing about the notion of taking themselves out of the expert's comfort zone and plunging headlong into the unknown territory of a new discipline' (Martin 2015). And the customer strategist Reuven Gorsht, blogging at Forbes, writes, 'Every company will need to consider the new capabilities they require to solve [their wicked problems]. [W]e will continue to see the rise of Generalists [as key figures in this problem solving]' (Gorsht 2013).

What these writers share is a belief that the appreciation of generalism or polymathy in both academia and outside has declined unduly in recent times; hyperspecialism, monodisciplinarity, and other synonyms indicating a narrowing of the mind and intellectual outlook must now give way more regularly so that, in Isaac Barrow's words, one may see the 'connection of things, and dependence of notions', for 'one part of learning doth confer light to another' (Barrow et al. 1700, p. 220).

The distinction between polymathy and generalism is an interesting topic for the history and sociology of knowledge (and also for the psychology of learning), but perhaps we need not be too concerned about it for the purposes of this chapter, at least. Our argument here is that wholesale academic specialization at undergraduate level, largely promoted and valued in recent times, is losing its worth in the knowledge economy, and, more positively, that something else can take its place. Both polymathy and generalism stand opposed to a narrowing of educational vision and worldview; they are united against monodisciplinarity as a useful and valid way to educate all students. We may educate in a modern version of polymathy or generalism both because such an education has an inherent, one might almost say traditional, value and because the wider changes we detect mean that such an education is likely to be the most useful one for future employment.

EXAMPLES FROM UCL'S ARTS AND SCIENCES BASc DEGREE PROGRAMME

To conclude: some examples of the student diets and graduate progression of UCL Arts and Sciences students. One student, who has been offered a prestigious graduate training contract with a City law firm, focused on politics, philosophy of science, and the study of cities, including more advanced spatial modelling and data mapping, which combination allowed her 'to offer [her] prospective employers a wider range of skills, and an ability to establish a good understanding with [legal] clients from the field' (personal communication; May 28, 2015). Another student has specialized in design architecture and psychology but, following an internship arranged by the Arts and Sciences programme, was offered a job at a financial consultancy firm in London. The firm appreciated her ability to create the workflow design of a financial restructuring programme. As an example of the eclecticism and creativity in many modern workplaces, this student commented that the firm was experimenting in part of its operation with 3-D printing, something she had covered on her arts and sciences course: 'The company was doing a side project 3D printing, and in my last term I had learnt how to 3D print…. you never know when such knowledge will be useful!' (UCL, 'What our students say about the BASc', 2015). A further student has been offered a job at a leading actuarial company after studying organic chemistry, management accounting, and Arabic. She describes the way in which the strategic approach to learning she gained on her degree transferred directly to her ability to impress at interview and during recruitment procedures.

Another of our graduates is employed at a global bank—interestingly, perhaps, straight from her undergraduate programme, without the need for a Master's degree; she comments:

> The Arts and Sciences BASc program is set up in such a way that really pushes the boundaries of education, teaching each of us skills as well as allowing us to explore various academic disciplines. It really makes you think outside the box and allows you to observe everything you do in a much broader context which I think helped me so much during my internship. So many people are very focused on one small thing that they forget where that fits in in the big picture, and I think that the Arts and Sciences degree really gets you into that frame of mind. I had an idea that I wanted to pursue a career in finance when I started out, and I thought that perhaps studying a slightly non-traditional degree may hinder my chances in such a competitive industry, but I think it did the opposite! The ability to customise your degree to your interests, means that each BASc student is unique and everyone brings new ideas and skills to the table—something every company wants. (UCL, "What our students say about the BASc," 2015)

Compare the thoughts of this student to the sociologist Steve Fuller's epigraph at the head of this essay. The student's words support the idea of a polymathic, liberal education and its usefulness in white-collar work.

A few more examples of educational profiles: we have development economists who combine engineering, economics, and policy courses and have written consultancy reports for law firms; a student studying materials chemistry alongside design and engineering; and students interested in digital humanities or combining neuroscience and literature. Some of these students are progressing to PhDs and Master's programmes in science, social science, or the humanities, and all are well-placed, if they choose, to take posts in parts of government, in the creative industries, marketing, law, development, finance, consultancy, media, etc.

At the time of writing, we have approaching one hundred examples of the progression of graduates and finalists, and this should increase by a hundred or so per year, now that the UCL course is established. It would be interesting to develop both more quantitative research on patterns of progression of graduates from Arts and Sciences programmes into the knowledge economy and finer detail on the sort of work these graduates are doing. Our impression is that white-collar graduate work is immensely varied, often contains a 'research' or 'creative' element (as business commentators John Hagel and others suggest) and that 'sector analysis', such as done in, for example, the UK Standard Industrial Classification of Economic Activities, is struggling to cope with the shifting boundaries and complexity of knowledge work. The example of the design and psychology student, offered a job and excellent starting salary at a financial consulting company, who ended up, by chance, contributing to the firm's '3-D printing project', is fairly typical in its variety and 'difficult-to-definedness'.

In some cases, the educational journey of young people educated in the liberal arts and sciences model may be one or two years longer than that of traditional specialists. Thomas Young and Da Vinci took more than a four-year college education to develop their talents: no one claims polymathy is easy! But with such an education, our young polymaths are likely to be better prepared for the creative challenges that await them in the knowledge economy.

This chapter has argued for the value of a liberal arts and sciences education as a good preparation for a large and growing number of jobs in the developed world. It has connected such an education both with some historic values of polymathy and generalism and some contemporary requirements of business and the burgeoning services sector. But of course one would not want to claim that *everyone* should be educated in this way. This would be to fall victim to the well-known fallacy of believing that one cannot have too much of a good thing! There will always be a need for specialist academic training. However, in the UK at present there is no danger of too many students being educated in a broad or polymathic way. A recent study estimates that less than 1% of current undergraduate programmes are of truly interdisciplinary nature (Au 2014). In China and India, too, in the late twentieth century there has been emphasis on specialization at under-graduate level. There is therefore room for polymathic arts and sciences programmes to grow. We will watch with interest how these global power-houses bring their cultural traditions to bear on polymathy and generalism, and the educating of the next generation of global graduates.

NOTES

1. By 'Inherent' vs 'instrumental' is meant here, roughly, the same distinction as between intrinsic and extrinsic value, i.e. the value something has as a 'good in itself' compared to the value something has as a 'means to an end'. See e.g. G.E. Moore's moral philosophy at http://plato.stanford.edu/entries/moore-moral/.

2. Of course, there is no guarantee that universities will always be regarded as necessary to produce graduates for white-collar work. All it may take is for five or six human resources departments from major graduate recruiters to announce that they will henceforth recruit differently for the system to suffer massive disruption. At the conference in Shanghai at which this paper was originally given, Jeffrey Lehman was right to point out that the ebbs and flows in graduate recruitment patterns now depend on an increasingly professionalized human resources sector. However, to date, employers in the UK at least still regard graduation certificates as necessary sieves in the recruitment process.

3. The mockumentary parodies '2012' and 'W1A', both produced by the BBC.

4. The rise of professional sport is just one such example but this is also true of jobs in public relations, much office work and, of course, almost anything associated with computers.

5. To be cautious and not deny the main point of Perkins' argument, namely the mass movement into white-collar work, we should probably call this the 3½ th revolution!

6. The Arts and Sciences BASc is a major new interdisciplinary liberal arts and sciences degree at UCL. It accepted students for the first time in 2012 and graduated the first cohort in 2105. It now accepts 120 students per year. Full details of the programme can be seen at www.ucl.ac.uk/basc.

7. The issue of the productivity of job titles is under review in some knowledge economies. I recently had the pleasure of confirming this with Mr Kass

Hussain, Director of Connected Homes at the major UK utility firm British Gas, who had been quoted by the BBC as saying that his business had found it productive to abolish job titles in parts of the organization; these titles limited people at a time when maximizing creative input from everyone was a priority.

8. The sector of manu-services which involves 'co-creating' products with customers is one novel aspect of a knowledge economy.

REFERENCES

Au, S. (2014). *Higher education in the United Kingdom.* Unpublished manuscript.

Barrow, I., Tillotson, J., & Hill, A. (1700). The works of the learned Isaac Barrow. Retrieved from https://play.google.com/store/books/details?id=ZmVZAAAAYAAJ&rdid=book-ZmVZAAAAYAAJ&rdot=1

Brown, P., Lauder, H., & Ashton, D. (2011). *The global auction.* New York: Oxford University Press.

Burke, P. (2010). The polymath: A cultural and social history of an intellectual species. In D. F. Smith & H. Philsooph (Eds.), *Explorations in cultural history: Essays for Peter McCaffery* (pp. 67–79). Aberdeen: Centre for Cultural History.

Chen, D. H. C., & Dahlman, C. J. (2005). The knowledge economy, the KAM methodology and World Bank operations [pdf]. Retrieved from http://siteresources.worldbank.org/KFDLP/Resources/KAM_Paper_WP.pdf

Deloitte. (2014, February 1). Value of connectivity. Retrieved from http://www2.deloitte.com/content/dam/Deloitte/ch/Documents/technology-media-telecommunications/2014_uk_tmt_value_of_connectivity_deloitte_switzerland.pdf

Deresiewicz, W. (2014, July 21). *Don't send your kids to the Ivy League.* New Republic. Retrieved from http://www.newrepublic.com/article/118747/ivy-league-schools-are-overrated-send-your-kids-elsewhere

Economist. (2014, January 18). The on-rushing wave. Retrieved from http://www.economist.com/news/briefing/21594264-previous-technological-innovation-has-always-delivered-more-long-run-employment-not-less

Ferris, T. (2007, September 14). The top 5 reasons to be a Jack of all trades [Blog post]. Retrieved from http://fourhourworkweek.com/2007/09/14/the-top-5-reasons-to-be-a-jack-of-all-trades/

Gombrich, C. (2013, June 22). The new amateurism [Blog post]. Retrieved from http://www.carlgombrich.org/the-new-amateurism/

Gombrich, C. (2015a, March 24). The knowledge economy and the end of the inherent vs instrumental value conflict in education [Blog post]. Retrieved from http://www.carlgombrich.org/the-knowledge-economy-and-end-of-the-inherent-vs-instrumental-value-conflict-in-education/

Gombrich, C. (2015b, June 16). Expertise [Blog post]. Retrieved from http://www.carlgombrich.org/expertise/

Gorsht, R. (2013, December 28). New problems, new approaches—The rise of the generalist. Retrieved from http://www.forbes.com/sites/sap/2013/12/28/new-problems-new-approaches-the-rise-of-the-generalist/0

Graeber, D. (2013, August 17). On the phenomenon of bullshit jobs. *Strike Magazine*. Retrieved from http://strikemag.org/bullshit-jobs/

Hagel, J. (2009, November 14). Pursuing passion [Blog post]. Retrieved from http://edgeperspectives.typepad.com/edge_perspectives/2009/11/pursuing-passion.html

Haskell, T. L. (2000). *The emergence of professional social science*. Baltimore: The John Hopkins University Press.

Jarausch, K. H. (2004). Graduation and careers. In W. Ruegg (Ed.), *A history of the university in Europe. Volume III. Universities in the nineteenth and early twentieth centuries (1800–1945)*. Cambridge: Cambridge University Press.

Jarche, H. (2013, October 9). The future of work is complex, implicit and intangible [Blog post]. Retrieved from http://jarche.com/2013/10/future-of-work-is-intangible/

Martin, R. (2015, January 6). The myopia of expertise [Blog post]. Retrieved from http://indalogenesis.com/2015/01/06/the-myopia-of-expertise/

Michaels, F. S. (2011). *Monoculture: How one story is changing everything*. Canada: Red Clover Press.

New York Times. (2015, April 21). China's economy puts new pressure on its lopsided jobs market. Retrieved from http://tinyurl.com/os89afg

Nussbaum, M. (2010). *Not for profit*. Princeton: Princeton University Press.

Office of National Statistics. (2013). UK service industries: Definition, classification and evolution [pdf]. http://tinyurl.com/pyl5a97

Open University. (2015). Careers. Retrieved from http://www.open.ac.uk/courses/qualifications/qd#careers

Perkins, H. (1996). *The third revolution: The rise of professional elites in the modern world*. London: Routledge.

Powell, W. W., & Snellman, K. (2004). The knowledge economy. *Annual Review of Sociology, 30*, 199–220.

Robertson, S. L. (2003). 'Producing' knowledge economies: The World Bank, the KAM, education and development. In M. Simons, M. Olssen, & M. Peters (Eds.), *Re-reading education policies: Studying the policy agenda of the 21st century*. Rotterdam: Sense Publishers.

Sanders, I., & Sloly, D. (2012). *Mashup: How to use your multiple skills to give you an edge, make money and be happier*. London: Kogan Page.

Schaffer, S. (2013). How disciplines look. In A. Barry & G. Born (Eds.), *Interdisciplinarity*. New York: Routledge.

Shumway, D. R., & Messer-Davidow, E. (1991). Disciplinarity: An introduction. *Poetics Today, 12*(2), 201–225.

Susskind, R. (2013). *Tomorrow's lawyers*. Oxford: Oxford University Press.

Susskind, D., & Susskind, R. (2015). *The future of the professions: How technology will transform the work of human experts*. Oxford: Oxford University Press.

The BPTI's Channel [Screen name]. (2012, February 8). Steve Fuller (University of Warwick, UK) on interdisciplinarity part 1 [Video file]. Retrieved from https://www.youtube.com/watch?v=LdI_GDUUGDE

Twigger, R. (2011). Polymathy forum > What does the word polymath mean to you? [Blogpost]. Retrieved from http://www.roberttwigger.com/polymathy-forum/post/1450300

Twigger, R. (2015, March 29). Polymathics [Blog post]. Retrieved from http://www.roberttwigger.com/polymathics/2015/3/29/polymathics-1.html

UCL. (2015). What our students say about the BASc. Available at: http://www.ucl.ac.uk/basc/comments

Weinberger, D. (2011). *Too big to know*. New York: Basic Books.

White, L., & Taket, A. (1994). The death of the expert. *The Journal of the Operational Research Society, 45*(7), 733–748.

Work Foundation. (2010). Knowledge economy strategy 2020 [pdf]. Retrieved from http://www.theworkfoundation.com/DownloadPublication/Report/263_CSR%20Submission%20FINAL%2030-9-2010.pdf

University College Freiburg: Toward a New Unity of Research and Teaching in Academia

Nicholas Eschenbruch, Hans-Joachim Gehrke, and Paul Sterzel

Abstract This contribution introduces University College Freiburg (UCF) and its English-taught Liberal Arts and Sciences (LAS) program. In the first part, the chapter relates the Freiburg LAS endeavor to the traditional continental European understanding of the unity of research and teaching. It then proposes five fundamental perspectives that the authors believe are particularly relevant for an emerging European model of liberal education: student autonomy and independence; an interdisciplinary and epistemological outlook; an early research orientation; the curricular integration of skills training; and an internationally oriented, multilingual mission within a larger university. In the second part, the LAS curriculum is introduced and situated in the context of UCF, a platform institution for interdisciplinary teaching at the University of Freiburg.

N. Eschenbruch (✉)
Centre for Security and Society, University of Freiburg, Freiburg, Germany

H.-J. Gehrke (✉) • P. Sterzel (✉)
University College Freiburg, University of Freiburg, Freiburg, Germany

© The Editor(s) (if applicable) and The Author(s) 2016   91
W.C. Kirby, M.C. van der Wende (eds.), *Experiences in Liberal Arts and Science Education from America, Europe, and Asia*,
DOI 10.1057/978-1-349-94892-5_7

Keywords Educational History • Germany • Teaching and Research •
Autonomy • Interdisciplinarity and Epistemology

From a German perspective, Liberal Arts and Sciences (LAS) study programs are a recent phenomenon, resting on ancient traditions of education
and more modern interpretations thereof, especially in North America, but
more recently also in the Netherlands. In the process of establishing a LAS
program in Germany, at the University of Freiburg, it has been both fascinating and challenging to reinvent German higher education while at the same
time finding—through reading the German classics, especially Wilhelm von
Humboldt (1982), closely—that a complete reinvention is hardly necessary.

The starting point for this chapter on the Freiburg LAS bachelor program
and its host institution, the newly founded University College Freiburg, is
our orientation toward personal education in the sense of German *Bildung*:
the education and formation of a person or personality—not merely the
training of specialized researchers, scientists, or scholars.[1] To reach this
goal, research-based formats are put to use—in the German tradition of
unifying research and teaching in higher education. As in research, this
takes the form of dealing with problems, topics, and objects as if they had
not yet been solved, discussed, or analyzed. The aim is not to reinvent the
wheel but to understand and convey how wheels are invented. Herein lies
the difference between mere reproduction and the creative acquisition of
knowledge that enables the education of the whole person.[2]

In this chapter, we will first look at some historical roots and general
considerations about science and its relevance for education. As a second step, we will then relate these general ideas to contemporary principles that characterize LAS programs in the framework of what one could
tentatively call an emerging European model, a model that was initiated
in the Netherlands about 15 years ago (van der Wende 2011). We will
then thirdly present the conclusions we have drawn from these observations and reflections in creating and implementing our LAS curriculum at
University College Freiburg.

SCIENCE AND EDUCATION

It was the ancient Greeks who "invented" the concept of academic training
by what they called the "arts of the free". Following on from Greek precedents, the Romans developed and formalized the *artes liberales* further.

During the Middle Ages, their concepts became an integral part of higher education and thus a central element in the foundation of European universities. This whole set of ideas then played an important role in the intellectual organization of modern universities since the Renaissance (see contributions to Zimmermann 2013 for more historical details).

The unity of science, scholarship, and teaching is, then, already reflected in the motto of the Collège de France, founded in 1530 by the French king François I in the spirit of European humanism: "enseigner la science en train de se faire" (teach science in the making). This overall goal was more recently reinforced by French philosopher Maurice Merleau-Ponty in his professorial inauguration speech in 1953: "Since its founding, the mission of the Collège de France is not to convey its audience acquired truths, but the idea of free research"[3] (Maurice Merleau-Ponty 1953, p. 9). The concept of "free thought" was deemed to be so important for and characteristic of the professors at the Collège de France that it was inscribed in golden letters above the institution's main hall.

From a German perspective, this coincides remarkably with the principles expressed by Wilhelm von Humboldt in his memorandum for the organization of higher education in Berlin (1810). He insisted on considering science and scholarship (in German, characteristically only one word: *Wissenschaft*) as "not a finished thing to be found, but something unfinished and perpetually sought after".[4]

As a working summary of the relevance of ancient ideas to modern practice, especially personal education, we arrive at six hypotheses:

1. Liberal Arts and Sciences education is not about reproducing knowledge—but about independence and the formation of an individual and independent judgment.
2. LAS education is not about faith in orthodoxies and ideologies—but about skepticism and questions that must be pursued by means of empirical research and experimentation.
3. LAS education is not about speculation—but about reflecting and thinking according to rational criteria, seeking formalization, and never disregarding the strict rules of logic.
4. LAS education is not about belief-based preaching—but about persuasive communication based on arguments and in accordance with the aforementioned rules.
5. LAS education is not about the quickest results—but about thorough reflection and honestly balanced argumentation.

6. LAS education is not about guarding arcane expert knowledge as an instrument of power—but about transparency and interaction with a broader audience.

Communication, diligence, capacity for reasoning, rationality, and skepticism—these qualities culminate in trained and rational judgment in the best and most comprehensive sense. They prove especially successful with regard to unsolved problems, indistinct topics, and unfamiliar objects of enquiry. This holds for "the unfinished and perpetually sought after" *Wissenschaft*, but is also valid for many areas of life beyond science and scholarship.

We do not mention these ideas for historical or traditional reasons or in order to suggest or encourage pure imitation. Rather, we believe that considering such principles is more urgent in higher education today than it has ever been in the history of academia. In the twenty-first century, we are faced with manifold and extensive global problems. Many seem to be nearly unsolvable or, to say the least, extremely complex. Approaching such problems must necessarily be based on rationality and thus especially on science, research, and scholarship: Essentially, this means that, in dealing with such problems and the general discourse about them, we need accurate observations and cogent arguments, and we need to educate people to develop that capacity.

Particularly when people, as agents, have responsibilities they have to be able to think and act accordingly. Educating people in this manner—that is, making use of the methods and patterns of research procedures—will enable them to lead their personal lives and to shape their professional activities as autonomously as possible, to acquire knowledge in the course of critical examination, and to act accordingly, in a self-reliant way, actively taking part in the self-rule of groups, institutions, nations, and the world.

Liberal Arts and Sciences Education: A European Model?

Moving on from the general historical precedents and approaches toward a more detailed programmatic understanding of the Liberal Arts and Sciences in a continental European higher education setting today, we arrive at another five points for discussion. The concepts and experiences in the Freiburg and German contexts do not (yet) carry the weight of the American tradition or the energy of Dutch innovation in the field. However, the success and reception so far allow for confidence and further initiative that another voice in this conversation will be heard. The following chapter traces out a space that accommodates an emerging German perspective on LAS, but is also open, with adaptations, to integrating academic traditions in countries that may not yet have institutional LAS settings.

Autonomy and Independence

In its varied interpretations since antiquity, the Liberal Arts and Sciences tradition has characteristically been seen as an education of students toward freedom. Their autonomy and independence, both regarding intellectual training and the personal formation that comes with it, have traditionally been emphasized and should be in the future. In this regard, there is much to be learned for continental Europe from North American "liberal education".[5]

Three aspects have been and still need to be addressed in particular: first, the inner freedom of character and intellect as a general disposition of the person. Second, education toward a purposeful use of outer freedom and choice in one's life. Coming from older philosophical roots, Isaiah Berlin (1969) famously laid out these two points for the English-speaking world. Third, the preparation for an active and responsible expression of freedom in a community and in relation to the common good. These connected aspects of freedom are still at the heart of Liberal Arts and Sciences education. In the German context, they resonate extremely well with Humboldt's idea of the centrality of "solitude and freedom" in academic education (*"Einsamkeit und Freiheit"*; Humboldt, 1810, p. 255).

Putting this orientation into practice can take on different forms: some LAS programs address it more implicitly, for example, with a highly choice-oriented curriculum. Others approach it explicitly, for example, by teaching core courses that self-reflectively discuss educational goals. In any case, an emphasis on freedom in some form should be a characteristic and defining part of any LAS curriculum. If it is to have credibility, it should also pervade institutional culture—especially teaching methods, student–teacher relations, and student involvement in academic self-governance.

Interdisciplinarity

A disciplinary approach toward teaching, already at the undergraduate level, has been central to university education in many European countries as well as a defining feature of German higher education since the nineteenth century. However, by forcing students to make an academic identity decision very early on, this model severely limits the exploration of freedom and autonomy as part of their intellectual education. It has also been argued that a strictly disciplinary approach is no longer adequate for contemporary challenges both in research and on the labor market.

In contrast, Liberal Arts and Sciences programs are interdisciplinary on a fundamental level. They encourage and require students to start university life by exploring the breadth of academic inquiry—humanities, natural sciences, and social sciences. In doing so, a particular feature of the LAS programs emerging in Europe is that interdisciplinarity does not exhaust itself in picking unrelated elements under the heading of general education. Rather, interdisciplinarity is didactically and intellectually integrated. This is typically achieved by both structured discussion of student choices—for example, through academic advising—and epistemological reflection in a core program.

Of course most LAS programs will still lead students to a specialization. In Germany, a large majority of students as yet pursue postgraduate, usually more specialized degrees. Consequently, LAS graduates can be expected to specialize during the course of their studies. However, from an LAS background, specialization can emerge out of disciplinary openness, through students' progressive, purposeful choices. In this sense, LAS programs as understood in this chapter can hardly be constituted by an interdisciplinary part tagged onto a fixed disciplinary curriculum or by an institution-wide core program implemented separately from the rest of teaching.

Research Orientation

Research universities have been a major European contribution to scholarship and learning worldwide. The European LAS programs that have emerged in the Netherlands, and later elsewhere, (van der Wende 2011) are situated in university cultures and higher education systems that traditionally favor an early orientation toward research and highly value research-oriented skills. LAS programs share the premise that their graduates need to be equipped with an understanding of research: not only as a formally necessary qualification for postgraduate study, but as a great way of training analytical, procedural, and personal competencies, and most of all because the adequate evaluation of research-derived information has become essential for many professional fields.

Thus, LAS programs of the type now emerging in Europe present and discuss knowledge itself throughout their curriculum, with reference to the most current research. They also familiarize students with general principles and theories of research and knowledge as such and train them in the procedural and presentational basics of the methodology of a student's chosen field of academic specialization. Finally, they challenge students to critically evaluate research results and knowledge claims in general.

As has been pointed out in the beginning, a research orientation need not be identical with actually doing research, and LAS programs do not train full-fledged researchers. This is reserved for postgraduate study, which also is important for a later, disciplinary or vocational sharpening of LAS graduates' professional profiles. However, beyond the educational benefits mentioned, a research orientation throughout the curriculum, as proposed here, clearly distinguishes LAS teaching and learning from school and general education. Furthermore, it reserves the inclusion into LAS programs for disciplines that actually do research.

Integrating Skills Training

In the course of the last decades, the conviction has been growing that there are important general professional and personal skills that are not automatically acquired in standard secondary education. The result has been an increasing number of university courses devoted to, for example, presentation skills, debate and public speaking, project management, or information technology skills. In Germany, these courses have characteristically been offered as optional courses or electives, often run by external trainers.

LAS programs have been quite open to the idea that not all general skills their graduates need are academic or scientific. However, they typically attempt not to teach such skills in isolation. Rather, they integrate them into the curriculum through specific teaching methods and learning environments. Problem-based learning, for example, fosters discussion and presentation skills. Project-based learning teaches students communicative competencies as well as project and time management. Acquiring these skills need not be separated from academic subjects, but should rather be discussed as an essential part of "Bildung", as it has traditionally been the case in Europe and recently been newly inspired by the Liberal Arts and Sciences model (for an overview of the German debates in this field in the wake of the Bologna reform see Spoun and Wunderlich 2005). Not every general professional skill a graduate might need can be covered by this approach. In fact, LAS programs are more likely to focus on generally applicable, person-related skills (speaking, presenting, writing, and dealing with numbers) that have traditionally been associated with academic and public life and can be practiced effectively in conjunction with academic learning. They are less likely to give credits for field-specific, nonacademic functional training.

International Orientation and Multilingualism

The academic world is undergoing intense internationalization. Therefore, and in recognition of English as the academic lingua franca, an English-taught curriculum can be seen as a precondition for successful internationally oriented study programs. This also means an explicit, ambitious institutional positioning in a global academic scene. In fact, a spirit of free enquiry, interdisciplinarity, and research orientation clearly cannot succeed without excellent knowledge of the most widespread academic language. While there are other European languages, notably French and German, that have eminent academic traditions and important contemporary scholarly and scientific communities, we argue that LAS programs must be connected to the global academic world through the full integration of academic English into their standard curriculum. To this end, language courses or exchange programs are not enough; a substantial part of the curriculum must be taught in English.

At the same time, however, LAS programs ought not only to honor, but also to continue non-English-based intellectual traditions by remaining open to and actively encouraging the acquisition and use of other languages. This is, of course, of particular significance in Freiburg, twenty kilometers from France and eighty from Switzerland, where multilingual universities have been a reality for a long time. The University of Freiburg is part of the international EUCOR network of Rhine Valley universities. It would be inadequate (and indeed unwelcome) to throw this overboard in favor of a university culture exclusively based on global English. In the spirit of freedom of enquiry and interdisciplinarity, an intelligent multilingualism that appreciates the best of the European scholarly traditions and bodies of knowledge, while at the same time further developing global connections, seems the appropriate solution. In this sense, by appealing to potential students from all over the world, LAS programs as outlined here could even be seen as the contemporary European contribution to the internationalization of higher education.

The Place of LAS Programs in Higher Education

In the Netherlands, LAS programs were explicitly introduced in order to initiate a vertical differentiation of the university system—and are well on their way of doing so (van der Wende 2012). It should be apparent that an emphasis on choice, interdisciplinarity, research orientation, high-level skills training, and multilingualism is not meant to address an unlimited number of students and—if understood correctly as a demanding personal challenge—may not appeal to large numbers either.

However, with view to other higher education contexts it may be premature to label or market LAS programs primarily as a new form of elite training: in current national university systems of many European countries, interdisciplinary study of this kind does not qualify for a number of important state-regulated professions like medicine or law. It will probably also remain a minority qualification in management and other important areas that traditionally promise social prestige and/or significant income. In a number of other fields, degree programs exist that are equally or more competitive to get in to than LAS.

In many European national contexts, LAS programs may thus be seen as a complementary and innovative element in higher education, rather than as a challenge to or replacement for existing models. In a space opened up by the Bologna process, they have the potential to address self-reliant students with a profile so far not catered to, answering emerging demands for responsible, academically trained flexibility, and internationality in the labor market, and doing so with a new vision of practice-oriented intellectuality beyond the traditional disciplinary order. While there are clear benchmarks for LAS programs, outlined above, there is also much room for individual programs in varied institutions to create their specific profile and for students to use these profiles for their own visions of education.

The Liberal Arts and Sciences Curriculum at University College Freiburg

In 2012, taking into account these general considerations and the current development of liberal education, the University of Freiburg established University College Freiburg (UCF). As a platform it is responsible for the organization of interdisciplinary and international teaching projects, mainly the Freiburg Liberal Arts and Sciences bachelor program. In conceptualizing this special curriculum, the University of Freiburg has adapted the abovementioned ideas to its own tradition and mission.

Building on its over 550-year history, the University of Freiburg has made "New Universitas" the guiding principle of its strategy for the future: this is characterized by a comprehensive range of academic fields ranging from theology to microsystems technology; by excellent competition results in research, teaching, and continuing education; and by a major international presence. In research, a focus on actively promoting interdisciplinary and transdisciplinary approaches in research centers has been established. However, although the University of Freiburg has been very successful in the field of teaching, its organizational principle remains the traditional disciplines and faculties. Particularly in view of Freiburg's New Universitas, the concept of "liberal education" is therefore a promising complementary approach. It sensitizes students to problem-oriented, interdisciplinary learning and research from the beginning of their studies; in so doing, it prepares them for the challenges of scholarship and professional life that can only be solved using the knowledge, methods, and skills of a range of disciplines. In a new form, this connects with the old educational traditions discussed above, such as the unity of research and teaching and the understanding of scholarship as "not a finished thing to be found, but something unfinished and perpetually sought after".

Liberal Arts and Sciences: Foundations

Freiburg's four-year, interdisciplinary degree program "Bachelor of Liberal Arts and Sciences" is conducted almost exclusively in English and offers students a broad interdisciplinary education, while at the same time providing for individualized academic concentrations on a high academic level. The program focuses on educating students in flexible interdisciplinary thinking and on teaching them to apply the problem-solving strategies they have acquired to complex interdisciplinary issues in scholarship and in practice. Within the program, students have greater choice than usual. This teaches them how to move forward in their academic and intellectual development in a focused, self-responsible, and independent manner— during their studies and beyond.

Thus, the program's foundations are oriented toward a model that has already been implemented successfully at several universities in the

Netherlands—mainly at Maastricht, Utrecht, and Amsterdam, with slightly different emphases in each place. In Freiburg as well, it was clear from the beginning that beyond the basic idea of an interdisciplinary curriculum, an LAS degree program should do justice to specific features and strengths of the University of Freiburg. Consequently, the interdepartmental working group responsible for designing the degree program defined three characteristic focus areas:

Knowledge

An epistemological focus ensures that, even in a wide-ranging program, the standard of scholarship is met and that students will acquire the intellectual tools necessary to connect their diverse learning experiences and develop their own epistemological convictions. Consequently, the origin, function, and contexts of knowledge are to be reflected across disciplinary boundaries, in both theoretical and applied respects, throughout the program. This approach connects LAS with the University of Freiburg's profile as a full-fledged research university and it extends this profile to include specific competencies for LAS students: they learn to consider problems from different perspectives, to be flexible in accepting new information, and to deal with uncertainty. The University has established two new professorships for the specific implementation of this area of the program: a Professorship of Epistemology and Theory of Knowledge and—a first in Germany—a Professorship of Science and Technology Studies.

Language and Culture

In Freiburg, unlike the USA and the Netherlands, the role of German as an academic language has to be taken into account. This challenge is addressed directly by a focus on language and culture, with intensive language education integrated into the curriculum. This ensures a high standard in academic English, brings all students who are not native German speakers up to a basic level in German as a requirement, and provides those who wish with the opportunity to master German at the academic level. The focus on language and culture also considers cultural differences in scholarship as well as in daily life and ensures that qualified graduates are able to use their excellent language skills in an expressive and culturally competent manner. This not only provides students with an additional qualification; it emphasizes that LAS in Freiburg stands for a diversity of content and scholarly traditions, and distinguishes the program from those offered solely in English.

Responsibility and Leadership

In designing the Liberal Arts and Sciences program, the University also surveyed the views of regional business associations. In discussions with these groups, the idea of adding a "responsibility and leadership" element to the LAS graduates' profile was developed. This enables the students to bring a high standard of leadership capability to their future professional responsibilities. Such capability is based on their intellectual flexibility (c.f. focus on understanding and knowledge) and training in effective communication and cultural sensitivity (c.f. focus on language and culture). As a result, LAS graduates are not just generalists with excellent communication and intellectual skills—they can also actively take responsibility and lead effectively in groups, projects, and organizations.

Liberal Arts and Sciences: Program Design

These three foundational perspectives—*Knowledge*, *Language and Culture*, and *Responsibility and Leadership*—are integrated in a concrete Liberal Arts and Sciences curriculum. The specific implementation of a "Core" section lays the groundwork for deliberate interdisciplinarity. Here, students furthermore receive intensive training in the cultural techniques that are necessary for effective creative activity in the professional world. This concentration constitutes the intellectual identity of the degree program. The actual focus of studies is then in the "Major": there, students choose a specialization in which they acquire a high level of academic competence in an academic field of their choice. Furthermore, students have a wide range of options for building individual focus areas within the "Electives" section, including the possibility to acquire and improve their foreign language skills.

The Core

The Core covers the theoretical, methodological, and practice-oriented education of all students. It shapes the identity of the program. In the first year of the program, the core combines five courses into a "First-Year Framework", which uses complex global challenges as points of departure for a problem-oriented introduction to the program. These topics—such as climate change or global justice—also provide a specific background for the application-oriented academic skills training. At the start of the program, the epistemological approach is taught on a problem-oriented basis. In the middle semesters, it is placed within a stricter philosophical framework in two highlighted seminars. These focus on the questions of what knowledge is, how it is created in the various branches of scholarship, and how it influences the world and is influenced by the world. These seminars are taught by the holders of the two new professorships that were established for the degree program. The later semesters of the program comprise the responsibility and leadership curriculum made up of four courses. This is intended to enable students to actively and effectively shape their environment and take responsibility for it. The epistemological aspect is still prominent here, but it is taught on a more application-oriented and experiential basis.

The Major

Every student in the Bachelor of Liberal Arts and Sciences degree program at UCF chooses a specialization (Major) as an academic focus area. Four Majors are implemented to date, and it is possible that others may be introduced. They are designed to remain interdisciplinary: they do not duplicate existing bachelor's degree programs at the University of Freiburg but instead enable new connections among fields and open up new perspectives. At the same time, they reflect academic fields that characterize the university's particular profile in research and teaching.

"Culture and History": This Major focuses on the cultural capabilities and forms of expression of human beings in contemporary and historical perspectives. It encompasses the fields of historical, literary, and cultural sciences as well as philosophy and art. The approach combines the traditions of various disciplines through their application to a broad conception of culture in history and the contemporary context.

"Life Sciences": This Major focuses on the bases of life from the biological perspective, connecting in particular with its capacity to be influenced by technology and its relevance for human thought and action. In this Major, therefore, the life sciences are not understood primarily in terms of molecular biology but are open to the technological sciences and biological anthropology.

"Governance": This Major focuses on issues related to the economic, legal, political, and administrative character and management of complex societies. Specifically, this field is considered not in terms of the differentiation of its disciplines but in terms of its functional interconnectedness.

"Earth and Environmental Sciences": From the perspective primarily of the natural sciences, this Major observes and analyzes global flows of materials, naturals systems, and their development in terms of both the earth's history and human impacts. It takes a systemic approach to understanding the planet, life, and the chemical interplay between physical and biological systems on the earth and in the environment.

The Electives

The Electives section gives students the opportunity to choose individual focus areas for their studies. They can select courses from the other LAS areas and enroll in appropriate courses in other degree programs in Freiburg. In addition, they have the opportunity to engage in supervised independent scholarly work, whether in cooperation with scholars at the university or at outside academic institutions, and to complete a professional internship or a practical project, over a period of up to three months, in order to obtain application-oriented experience before even completing their degree. Moreover, a practical project may also comprise an independent research project or a project with a social or artistic orientation that is in line with the program's educational standards. While the courses in Advanced Academic English are part of the Core section, students can use the Electives to meet the program's language requirements and to acquire or improve foreign language skills as an integral component of the degree program. The university considers it especially important for those students who enter LAS with no prior knowledge of German to have the opportunity, during their studies, to acquire sufficient German skills in order to take Electives taught in German in the later semesters and to pursue a German-language master's program after successful completion of their bachelor's degree.

The Liberal Arts and Sciences degree program is by no means intended to replace disciplinary degree programs at the University of Freiburg. The University of Freiburg sees Liberal Arts and Sciences as a meaningful, forward-looking addition to its educational profile. At the same time, it is a significant step in enhancing and expanding the university's international appeal in undergraduate education.

University College Freiburg

Organizationally, the Freiburg Liberal Arts and Sciences program is based in a newly established University College, a cross-faculty institution at the University of Freiburg. There are a number of political, educational, and pragmatic reasons for this embedded institutional setting: From a political perspective, it signifies the new institution's full commitment to the model of the nonresidential, publicly funded, comprehensive German research university. UCF was founded as a complementary element of such a university, not in opposition to it, and its founders believe this to be a general recommendation for LAS programs in Germany.

In this vein, UCF does not just stand for a decontextualized institutional model that could be implemented anywhere: it signifies one move forward in German higher education as a whole. A research university setting is in itself an invaluable framework for UCF and LAS. The unity of research and teaching, for example, is much more easily achieved in an institution that already hosts world-class research and researchers across subjects. Furthermore, a fruitful tension between disciplinary and interdisciplinary undergraduate education, between a German and an international academic outlook, can easily be generated when UCF students meet their colleagues from other programs, in their extracurricular life, and in some elective, joint classes. Finally, on a pragmatic level, an existing world-class university can provide resources of many kinds—reputation, legal expertise, famous alumni, broad teaching and research facilities, to name but a few—that help a new college to a quick and efficient start. And it should be mentioned that, in the German context, an integrated college is a lot more cost-efficient than building an institution from scratch.

In its role as an interdisciplinary educational center, UCF is a change agent that in turn will benefit the university in a number of ways. Most of all, in addition to providing an organizational framework for LAS, the university is pursuing three fundamental goals with the foundation of UCF:

First, UCF meets the needs of students, both in research and in the job market, for interdisciplinary forms of undergraduate university education. This is especially the case concerning the LAS program, but UCF also provides infrastructure for corresponding projects—for example, the project "Fascination of Science", in which students from different degree programs form interdisciplinary research teams and learn how a scholarly approach works in practice. The starting point is a topic that can only be addressed adequately from an interdisciplinary perspective (e.g. "Water scarcity in the Middle East" or "Sleep"). After appropriate subject-matter study, the module leads to the development of independent research projects.

Second, UCF serves as a teaching laboratory for innovative forms of university teaching. At UCF, instructors from throughout the university have scope to investigate new forms of teaching and then transfer their experiences to their respective departments. Conversely, UCF provides a university-wide platform for the numerous innovative approaches that already exist in the area of teaching, further enhancing their visibility. It is intended to open LAS courses to appropriately qualified students in other programs. In this way, UCF will integrate instructors and students from the entire university and will enable interdepartmental exchange in the area of innovative university teaching. In the future, UCF will also provide planning and organizational support to the existing international, interdisciplinary master's programs at the University of Freiburg.

Finally, as a third objective, University College Freiburg seeks to make a substantial contribution to internationalization in the area of teaching. UCF appeals to new student groups from abroad, has established a broad range of international partnerships and exchange programs, and thereby raises the university's global profile even further.

UCF thus provides coordination and support for the projects based under its roof and it enhances the work of the faculties and departments in the area of innovative interdisciplinary and internationally oriented university teaching. In this way, it has an effect on the entire University above and beyond its core activities and will contribute to raising the profile of and further expanding the interdisciplinary and international approach and outstanding quality of teaching at the University of Freiburg.

Outlook

In 2008, the University of Freiburg first considered establishing a University College and a Liberal Arts and Sciences degree program based there. In view of the general challenges for teaching in the current university environment, it was a welcome and timely development for the university when the German Council of Sciences and Humanities (*Wissenschaftsrat*) made very similar proposals in its recommendations on differentiation of the university system in 2010. In these proposals, the Council advocated the establishment of University Colleges, based in particular on the Dutch model, and the establishment of interdisciplinary degree programs for suitably motivated students to expand the undergraduate program offerings (German Council of Sciences and Humanities 2010). Thanks to early planning, the University of Freiburg is the first full research university in Germany to have established an English-language degree program in Liberal Arts and Sciences.

University College Freiburg and its Liberal Arts and Sciences program are young institutions—especially against the backdrop of an *artes liberales* tradition that is over two millennia and a university that is over 500-years old. As we made clear at the beginning, it is not our intention to either discount or glorify historical precedents and developments in designing an undergraduate education for the twenty-first century. Rather, the history of higher education offers an archive of educational principles, experiences, and models of decisive relevance for our contemporary knowledge-based society. Liberal education models in place today that focus on educating independent and responsible young people who can work and act in interdisciplinary and intercultural contexts are, in our view, just as innovative and necessary training grounds as the "arts of the free" in ancient Greece. UCF and interdisciplinary undergraduate teaching are novel institutions in German higher education. The first four LAS classes have already been enrolled, each with about eighty students. At present, preparations are underway for the graduation of the first LAS class whose achievements we will celebrate in summer 2016. The national and international attention and positive feedback from colleagues, the general higher education institutions, the political level, and especially from our students prove that this was a successful and necessary decision and that we are heading in the right direction.

Notes

1. The German concept *Bildung* encompasses a range of meanings including education, formation, (self-) cultivation, or (self-) development (Bruford 1975). For the purpose of this chapter, the English word education is used.
2. For a recent discussion (Figal 2013).
3. "Ce que le Collège de France, depuis sa fondation, est chargé de donner à ses auditeurs, ce ne sont pas des vérités acquises, c'est l'idée d'une recherche libre."
4. Translation: Ash 2006, p. 246; original: "Alles beruht darauf, das Princip zu erhalten, die Wissenschaft als etwas noch nicht ganz Gefundenes und nie ganz Aufzufindendes zu betrachten, und unablässig sie als solche zu suchen." (von Humboldt 1982, p. 257).
5. For a look at that tradition from a German LAS perspective see Plumley (2013).

References

Ash, M. G. (2006). Bachelor of what, master of whom? The Humboldt myth and historical transformations of higher education in German-speaking Europe and the US. *European Journal of Education, 41*(2), 245–267.

Berlin, I. (1969). Two concepts of liberty. In I. Berlin (Ed.), *Four essays on liberty.* Oxford: Oxford University Press.

Bruford, W. H. (1975). *The German tradition of self-cultivation: "Bildung" from Humboldt to Thomas Mann*. Cambridge: Cambridge University Press.

Figal, G. (2013). Die Bildung und Humboldt. In B. Zimmermann (Ed.), *Septem: Von Artes Liberales zu Liberal Arts* (pp. 109–118). Freiburg: Rombach.

German Council of Sciences and Humanities. (2010). Recommendations on the differentiation of higher education institutions. Retrieved from http://www.wissenschaftsrat.de/download/archiv/10387-10_engl.pdf. Accessed 12 Oct 2015.

Merleau-Ponty, M. (1953). *Eloge de la philosophie et autres* essais [In praise of philosophy and other essays]. Paris: Les Éditions Gallimard.

Plumley, R. (2013). Liberal Arts als Bildungsideal in den Vereinigten Staaten [Liberal arts as an education ideal in the United States]. In B. Zimmermann (Ed.), *Septem: Von Artes Liberales zu Liberal Arts* (pp. 119–138). Freiburg: Rombach.

Spoun, S., & Wunderlich, W. (Ed.). (2005). *Studienziel Persönlichkeit. Beiträge zum Bildungsauftrag der Universität heute* [Personality as study goal: Contributions towards the educational mission of the university today]. Frankfurt/Main: Campus.

van der Wende, M. C. (2011). The emergence of liberal arts and sciences education in Europe: A comparative perspective. *Higher Education Policy, 24*, 233–253.

van der Wende, M. C. (2012). *Trends towards global excellence in undergraduate education: Taking the liberal arts experience into the 21st century* (Research & occasional paper series: CSHE, 18). Center for Studies in Higher Education, University of California, Berkeley.

von Humboldt, W. (1982 [1810]). Über die innere und äussere Organisation der wissenschaftlichen Anstalten zu Berlin [On the inner and outer organisation of the Academic Institutes in Berlin]. In A. Flitner & K. Giel (Ed.), *Wilhelm von Humboldt: Werke in fünf Bänden* [Works in five volumes] (Vol. 4: Schriften zur Politik und zum Bildungswesen) (pp. 255–266). Stuttgart/Darmstadt: Wissenschaftliche Buchgesellschaft.

Zimmermann, B. (Ed.). (2013). *Septem: Von Artes Liberales zu Liberal Arts*. Freiburg: Rombach.

An earlier version of some parts of this paper (pp. 99–104) has been published in German as

Liberal Arts and Sciences an der Volluniversität des 21. Jahrhunderts—das Freiburger Modell, [Liberal arts and sciences at a 21st century comprehensive research university—the Freiburg Model] in: *Vigonianae, 2*, 59–65.

CHAPTER 8

The Liberal Arts and the University: Lessons for China in the History of Undergraduate Education in the USA and at the University of California

Nicholas B. Dirks

Abstract In this chapter, University of California (UC) Berkeley Chancellor Nicholas B. Dirks traces the origins of undergraduate education in the USA and at the UC, reviewing the liberal arts tradition in the context of the system of public higher education that developed in the USA. Through this chapter, Dirks argues that US public higher education and its blend of liberal learning in tension with pre-professional training has been immensely important to the nation, and is worth not only preserving, but also enhancing and strengthening, given its contributions to economic growth, innovation, socioeconomic mobility, civic engagement, and cultural vitality. He ends by articulating the growing importance of a liberal education to preparing students around the world for the challenges of life in the twenty-first century.

This chapter is based on a presentation delivered to the UC Regents on March 18, 2015. Sincere thanks to John Aubrey Douglass, Carol Christ, Nils Gilman, Dan Mogulof, and Michael Emerson Dirda for their role in preparing this chapter and for their valuable insights, contributions, and comments.

N.B. Dirks (✉)
University of California (UC), Berkeley, CA, USA

W.C. Kirby, M.C. van der Wende (eds.), *Experiences in Liberal Arts and Science Education from America, Europe, and Asia*,
DOI 10.1057/978-1-349-94892-5_8

Keywords University of California • History of Undergraduate Education • California Master Plan • Globalization of Higher Education

INTRODUCTION

How does higher education benefit society? How does it benefit students? What should be taught in college? What should a university look like? Who should attend it? Who should fund it? Who should govern it? The answers to such fundamental questions have never been fixed, but examining the continuing evolution of those answers is essential to both understanding contemporary conceptions of the university and to proposing future structures and needs.

This chapter examines these questions, and how they have been addressed over time, by tracing the origins of undergraduate education in the USA and at the University of California (UC), and reviewing the liberal arts tradition in the context of the system of public higher education that developed in the USA. Through this historical sketch, I hope to show that—contrary to claims that colleges and universities are "the slowest-changing institutions in American life,"—our educational institutions have indeed been remade many times over in response to shifting social, economic, and cultural currents (Delbanco 2014, p. 22). Adapting to and meeting changing societal needs, indeed, has allowed the system of higher education that emerged in the USA to be continually relevant and to contribute prodigiously to economic growth, innovation, socioeconomic mobility, civic engagement, and cultural vitality in our nation.

China and other countries now moving from "elite" to mass higher education can learn much from American public higher education, its interweaving and tiered systems, and its unique blend of liberal learning in concert (and occasional tension) with pre-professional training. That said, ours is by no means a perfect model, and indeed there are areas that our system has neither settled for good nor even, in some instances, begun to address in serious ways. Most pressing among these is sufficient adaptation to globalization. Top American universities all have substantial numbers of foreign students, offer a growing number of courses in a wide range of international subjects, support a broad spectrum of study abroad programs, and collaborate in an expanding array of research with foreign partners. But we have only started to come to terms with the volume and velocity of global connections, and have not gone nearly far enough in altering our content and methods to support students in a deeply interdependent world.

When planet-wide problems do not recognize either national borders or the boundaries that have traditionally separated academic disciplines, universities must adapt. Any burgeoning university system, too, should take advantage of the opportunity to build around this critical aspect of modern life.

Before turning to such contemporary issues, however, let us first examine the past of undergraduate education in the West.

CLASSICAL AND RELIGIOUS EDUCATION

Like much of Western society, our systems of higher learning have roots in the classical thought of ancient Greece. What we think of as formal education is largely based in the structured, systematic study of a body of knowledge as promoted by the great classical philosophers Socrates, Plato, and especially Aristotle. Those thinkers were concerned with fundamental questions about existence and human nature—What is being? What is truth? What is wisdom? What is virtue? What is good?—and their followers developed a method of studying those questions that took the form of seven essential topic areas, divided into two categories: the trivium, consisting of the verbal arts of logic, grammar, and rhetoric; and the quadrivium, consisting of the numerical arts of mathematics, geometry, music, and astronomy.

Beyond serving as a basis for metaphysical inquiry, these areas—the original "liberal" (meaning, "worthy of a free person") arts—were the subjects considered essential for citizens in ancient Greece in order to take an active part in civic life. At the time, this meant participating in public debate on the issues of the day, defending oneself in court, serving on juries, and serving the state through military service.

These seven subject areas provided a basic structure for intellectual life in early medieval universities as they emerged in the eleventh through thirteenth centuries. The university itself began as a congregation of people—the word *universitas* means simply a number of persons united into one body—not a physical place. Initially, meetings of the *universitas* took place where space was available, and did not have dedicated facilities unless a funder, often the church, provided one.

From medieval times through the Enlightenment, Christianity was integrally connected to Western universities, including all of those established in America before the revolution. Members of the clergy taught classes, and the small number of students who attended colleges predominantly sought positions as ministers later in life. Religious study ruled much of student life: chapel was held each morning and evening, and prayer and the study of scripture were major components of an education.

While a curricular focus on the classical liberal arts began in this era, it was also framed by an early tension between reconciling the thoughts of antiquity, especially ideas related to understanding the natural world and our place in it, and those of the church.

Early theologians had posited that the world was composed of ideas in the mind of the divine, which man could only "know" in an imperfect, mortal sense. But later ones, notably Thomas Aquinas, believed that knowledge and understanding of these ideas and their purposes could be deduced systematically through logical means. While students benefited from memorization and recitation of scripture, then, they could come to greater spiritual understanding through the study of the laws of nature, laws that were both divine and logical. This purpose of education was made explicit in universities' missions; at its founding in 1636, for instance, Harvard's original student handbook stated, "the main end of the student's life and studies is to *know* God and Jesus Christ."

Such logical analyses were an essential part of learning, but the other two components of education in early universities were the mastery of languages—Latin for instruction, Greek to read the New Testament, and Hebrew to read and translate the Psalms and the Old Testament—as well as formal scholastic debate about religion and the moral subject.

While retaining significant ties to the church, universities began to shift their priorities during the Enlightenment. Education came to be seen as key to the development of "gentlemen"—men who had inner virtue and outward manners; who understood honor, generosity, independence, and fidelity. Professions such as medicine and law could exist outside of college, so an education was sought after for personal, not professional, betterment. At the same time, education began to be seen as a requirement for the functioning of the polity and civil society, especially for a new nation.

HIGHER EDUCATION AT THE DAWN OF THE REPUBLIC

Immersed as they were in classical history and thought, the founders of the USA assumed that the survival of republics hinged on their citizens' abilities to put the public good—*res publica*—above personal interests. At the dawn of the republic, universities developed a new institutional focus and altered curricula designed to forge citizens who would strengthen the burgeoning nation as engaged and responsible participants in a democracy.

At the time of the American Revolution, all but one of the nine colleges then in existence in North America supported independence from Britain.

The colleges were hubs of sedition, providing intellectual training to young leaders of the Revolution and converting the ambivalent into supporters of independence (Tucker 1979, p. 18). Significantly, revolutionary fervor caused a spike of interest in history, governance, political theory, and law. While the proportion of graduates entering the church dropped from one third in the 1760s to one fifth in the 1790s, graduates who became lawyers jumped from 13% to 30% (Geiger 2014, p. 145). Latin writings detailing the fall of the Roman Republic and the beginning of the Empire were increasingly taught; and classic works evaluating systems of government or focused on civic morality—such as Cicero's orations, Caesar's Commentaries, and Tacitus' histories—took new precedence as well (Robson 1985, p. 166). The political lessons seemed clear: as Yale President Ezra Stiles noted in 1777, "it is scarcely possible to enslave a republic where the body of the people are civilians, well instructed in their laws, rights, and liberties."[1]

Most political leaders at this time did not consider it possible to educate everyone, but many did see value in producing what Thomas Jefferson called the "natural aristocracy" of learning and talent not tied to social class. Envisioning universities as essential to turning America into a true meritocracy—and in backlash against class-conscious European society—Jefferson developed the first notion of university education as a means of achieving social mobility. In founding the University of Virginia in 1819, he created the first publicly supported college, dedicated to educating leaders in practical affairs and public service rather than for either the pulpit or the professions. It was the first university without a religious affiliation to be established in the USA.

SCIENTIFIC AND PRACTICAL EDUCATION

After the Revolutionary War, as American society became more industrial, new practical and vocational interests also altered the curricular focus of universities. Navigation, engineering, and mechanics were added to religious and moral training. The study of classical languages began to give way to modern languages such as French and German.

Since Newton, basic science had a secure place in the American curriculum, with physics and astronomy held in particularly high regard. Developments in the industrial era, though—the railroad boom and an attendant need for civil engineers, for example, and the 1840 publication of Justus von Liebig's *Organic Chemistry in its Application to Agriculture and Physiology,* identifying the roles of nitrogen and minerals in plants and explaining the mechanism behind fertilizers—provided displays of scientific knowledge tangibly serving economic interests.

As recognition spread that universities could offer practical education benefitting the economy, some pre-revolutionary era colleges were accused of failing to be—in the words of Amherst College's 1827 charter—"sufficiently modern and comprehensive, to meet the exigencies of the age and the country" (Packard 1827).

Critiques like these led to a curriculum review at Yale after the college's trustees advocated dropping the study of "dead ideas and languages." President Jeremiah Day responded by noting how the curriculum in fact had evolved to include such new areas as trigonometry, surveying, and mineralogy, but also that the purpose of college was "not to finish a preparation for business, but to impart that various and general knowledge which will improve, and elevate, and adorn any occupation." Day's belief that college should "lay the foundation of a superior education" through "the discipline and furniture of the mind; expanding its powers and storing it with knowledge" served as a mandate for the liberal arts, and his so-called Yale Reports of 1828 lasted in influence through much of the nineteenth century, and in some respect to the present day, even when the specific curricular recommendations they contained were most disputed (Yale College 1828).

The new conceptions of the university introduced in the early to mid-nineteenth century were often added to existing ones. Indeed, the university as a means to produce skilled labor and expand human knowledge accepted the basic tenets of traditional liberal study, if in the context of some debate over what kind of knowledge might be useful, even granting the foundational need for a moral sensibility. This notion that colleges could both turn out more useful citizens and generate useful knowledge was the basic idea framing the development of a national system of public colleges, the land grant universities.[2]

JUSTIN SMITH MORRILL AND THE LAND GRANT ACT

On July 2, 1862, President Lincoln signed the Land Grant Act, giving states 30,000 acres of federal land for each of their Congressional representatives and senators, to be used to establish an endowment supporting:

At least one college where the leading object shall be, without excluding other scientific and classical studies, and including military tactics, to teach such branches of learning as are related to agriculture and the

mechanic arts, in such manner as the legislatures of the states may respec-
tively prescribe, in order to promote the liberal and practical education of
the industrial classes on the several pursuits and professions in life (First
Morrill Act 1862).

Authored by Vermont representative Justin Smith Morrill, the Land
Grant Act was not groundbreaking in theory—land grants had been
already used to support education in the 1787 Northwest Ordinance,
practical education was a stated priority for many universities, and agri-
cultural colleges were even then operating in Ohio, Michigan, and elsewhere. What broke new ground, however, was the scale of the new act,
enabled by Morrill's political acumen and the specific formula he adopted.
Choosing to weight a state's grant by its number of Congressmen (and
thus state population) made the idea attractive to skeptical Easterners,
and requiring that proceeds be used only as endowment induced states to
make a permanent commitment to their colleges.

In all, sixty-nine colleges were established or expanded through the
act—including, in 1868, the University of California. In envisioning a comprehensive system of colleges for the industrial class—the "thousand willing
and expecting to work their way through the world by the sweat of their
brow" (Missouri State Board 1866)—Morrill elevated the practical voca-
tions of agriculture and mechanics to the same social standing as the liberal
arts and sciences, while ensuring that any citizen could have access to both.

RESEARCH AND SPECIALIZATION

In the late nineteenth century, Henry Tappan, the first president of the
University of Michigan, articulated the idea of the university as a place
devoted not only to conveying existing knowledge, but also rooted in
discovering new knowledge. He defined universities as:

> Cyclopedias of education: where, in libraries, cabinets, apparatus, and pro-
> fessors, provision is made for studying every branch of knowledge in full, for
> carrying forward all scientific investigation; where study may be extended
> without limit; where the mind may be cultivated... in the lofty enthusiasm
> of growing knowledge and ripening scholarship. (Tappan 1851, p. 68)

At the turn of the nineteenth century and increasingly throughout the
twentieth, the vision of the comprehensive university based in research
activity for the advancement of knowledge was seen as foundational for

public as well as private universities, led in particular by Michigan, Johns Hopkins, Chicago, and Columbia. Building off of a new German model of the university developed by Wilhelm von Humboldt, American institutions began to invite students into the process of the discovery of knowledge, "encouraging productive thinking" rather than the "regurgitation of knowledge" (Röhrs 1987, p. 20).

This development had significant impact on undergraduate education. As the research foci of the university expanded, the variety of classes and specialized areas of study increased greatly. The formation of academic societies and the hiring of specialist faculty led to a proliferation of fields of study: universities hired professors of Sanskrit, Arabic, and Chinese; they developed new fields in social science that evaluated concrete social issues including prison reform, poor relief, crime, and deviance. At the same time, the pillars of the traditional undergraduate education—curricular separation of individual colleges, fixed courses for the bachelor's degree, required Latin and Greek, routines structured around recitation—began to be replaced by curricula based on individual academic disciplines and increased student choice.

The idea of a specialized focus in a major, especially organized by academic discipline, only developed slowly during the nineteenth century. While the University of Virginia allowed students some freedom in shaping their studies early on, the first use of the distinction "major" related to undergraduate degrees was at Johns Hopkins in 1877, and was an innovation directly connected to the university's commitment to research in the German mode.

Not long afterwards, at Cornell and Harvard, new elective systems granted students the freedom to explore personalized courses of study. Traditionally, classes had been attached to class year and named for texts to be covered. President Charles Eliot of Harvard reorganized courses by department, number, and instructor, and opened them up to all qualified students regardless of their year, championing what in 1885 he called a "spontaneous diversity of choice" (Harvard University 1885).

In 1905, the UC followed suit, dividing the curriculum into a lower and upper division. UC created a framework for liberal study in the first two years of education, and increased specialization in the latter two, introducing the major as the organizing principle for these years. Since then, the idea that undergraduate learning should include general education as well as specialization—typically in the form of the major—has been a distinguishing feature of American higher education.

At the turn of the century, while academic study remained central to the university, the collegiate experience began to be viewed more holistically, encompassing as it did valuable nonacademic components in addition to a classroom experience, acknowledging too the role of college in mediating youth and adulthood. Inspired in part by Teddy Roosevelt's call for a "strenuous life," athletics, especially football, became important to cultivating the "whole man." Fraternities and sororities grew and acquired chapter houses, which had the effect of attracting alumni back to campuses after graduation. Residential housing, student newspapers, the YMCA, glee clubs, and many other activities rounded out the collegiate experience. UC inaugurated the idea of the residential college in the establishment of Bowles Hall in 1927, although other colleges, most notably Yale and Harvard, developed complete systems of residential colleges soon thereafter. Universities' new focus on organizing these elements of the life of collegians underscored the growing acceptance that the value of college was not limited to what was taught in classrooms by professors.

ORGANIZING MASS HIGHER EDUCATION

During the first half of the twentieth century, the influence and importance of universities expanded not just for the elite but also for the growing middle class. The President's Commission on Higher Education stated in 1947 that "every American should be enabled and encouraged to carry his education, formal and informal, as far as his native capacities permit," (President's Commission on Higher Education, vol. 1, p. 101) and it was in this era that many Americans began to see the university as necessary for personal fulfillment, economic betterment, and social success. This promise of access to higher education as a universal right was made explicit in acts like the GI Bill of 1944,[3] which made clear to returning veterans that college was the path to rejoining society and to having a prosperous life.

For UC, as for some other public universities, this meant balancing excellence in instruction with a need to vastly increase capacity. Early in the twentieth century, as UC expanded enrollment, its faculty worked with public high schools to review curricula and set standards that would enable students to thrive at the college level. In 1907, the California legislature passed the nation's first bill to establish junior colleges as extensions of such high schools. Both students and businesses benefitted from the local, low-cost schools, which provided training for a growing white-collar labor force as well as the more advanced technical jobs in the blue-collar sphere. This reflected continuing disagreement about whether to create undergraduate degrees that would be exclusively professional, how to use professional schools and degrees for undergraduates, and what the meaning, reach, and significance of the liberal arts should be for the general population. It also set the stage for what would become, after World War II, the multi-tier, functionally differentiated system of higher education institutions that was a cornerstone of the 1960 California Master Plan for Higher Education.[4]

The California Master Plan was spearheaded by then UC President Clark Kerr but devised by a survey team appointed by the UC Regents and the State Board of Education during the administration of Governor Pat Brown. The plan formalized an interworking system of postsecondary education that gave specific roles to the UC, to the descendants of California's normal colleges or teaching schools, and to the state's community colleges. It associated a general commitment to the liberal arts with the research work of the top tier of the university system, while accommodating and serving a rapidly increasing population in need of new skills and advanced training across a multitude of fields. Under the banner of the idea of meritocracy, it provided the basis for the public support of elite higher education—the foundation on which the UC campus in Berkeley could be the peer of Harvard and indeed any other world university, private or public.[5]

To Kerr, the university had become a "prime instrument of national purpose." He argued that the knowledge produced at universities had become the main fuel for the growth of a nation, its military might, economic competitiveness, artistic excellence, societal contentedness, and political stability. In his classic book, *The Uses of the University* (1963), he wrote that, "What the railroads did for the second half of the last century and the automobile for the first half of this century may be done for the second half of this century by the knowledge industry." (p. 63).

Though students accused him during the 1960s of championing the corporatization of the university, Kerr was describing a new reality about which he had great optimism but also abiding concerns. Although he believed that the university would lead the way to new economic possibilities, he was well aware that it risked becoming a knowledge "factory" whose neglect of students through large classes, the overuse of teaching assistants, and the selection of faculty members based on their research expertise alone could alienate the undergraduate student body.

Much has been done over the subsequent fifty years across the UC to address these concerns, from the establishment of the visionary college systems in San Diego and Santa Cruz to the investment of huge resources on all of the UC campuses in student support, advising, housing, career counsel and planning, and perhaps most importantly, teaching. The Carnegie Foundation's 1998 Boyer Report,[6] in particular, prompted a nationwide discussion on how to better engage undergraduate students at major research universities, and led to the elevation of teaching in faculty advancement reviews and to the expansion of student involvement in faculty research. Other changes have further altered undergraduate education itself: Technology has and continues to change the way students learn, interact, and experience a modern liberal arts education, offering real and virtual learning environments that alter how they engage with peers, faculty, staff, and the university resources at their disposal. Inquiry-based learning, interdisciplinary opportunities, collaborative problem solving; the notions of global citizenry, ethics, and personal responsibilities; new models for mentoring—all of these, and more, form the foundation of an undergraduate education that is holistic in nature and also caters to the individual interests and abilities of students who come from increasingly diverse socioeconomic and ethnic backgrounds.

NURTURING THE FUTURE OF UNDERGRADUATE EDUCATION

What emerges in this brief historical sketch is that undergraduate education is constantly evolving, becoming increasingly complex and sophisticated in a manner that reflects the growth in knowledge about teaching and learning, the needs and desires of society, and the history of faculty investment in the fundamental purposes of the bachelor's degree. Even in our current era of state disinvestment from public higher education, the UC Berkeley, is at the forefront of efforts to redefine and rearticulate the centrality of undergraduate education and the liberal arts tradition not just for our teaching mission, but for the other domains in which we excel, namely research and public service. Indeed, we are becoming more committed than ever before to supporting the student experience in and outside of the classroom, as we seek to prepare students for the growing challenges of life in the twenty-first century.

Today, students here and abroad face difficult and sometimes daunting prospects in an economy where traditional jobs are shrinking and changing at a faster pace than ever before (students graduating today face a vastly different horizon of employment than ever before, and will have an average of at least six different kinds of jobs throughout their lives). This has led to some skepticism about the liberal arts, both as the term refers to majors in humanistic fields, and as general education for the purpose of a generic set of educational values. At Berkeley, we confront these debates by embracing change and innovation while also holding on to some traditions and values that have been for years a critical feature of the unique accomplishment that our university represents. We are committed to preparing our students to be able to reinvent themselves intellectually and professionally numerous times over the course of their lives. We believe that now more than ever, the liberal arts will play a critical role in the cultivation of this adaptive and creative capacity, even as we believe that in order to train future leaders, we must be especially attentive to critical thinking, general intellectual acuity, quantitative capacity, and civic preparedness.[7]

For China, where modern educational institutions developed out of deep interaction with institutions in the USA, Europe, and the Soviet Union, this recent history has profound relevance as well. China's emergence as one of the largest world economies only underscores the extent to which US debates about the liberal arts should influence the extraordinary investment the Chinese state is making in higher education. Not only is China experiencing rapid change in its workforce needs, it plays a critical role in world politics and as such must draw on appropriate Western models as it continues to assert a leadership role in the world.

As China's global footprint expands and as its higher educational system moves from elite to mass education, Chinese universities need to find effective ways to adjust their educational goals to cultivate a new generation of students who are creative, adaptive, and critical thinkers schooled in issues that range from morality and ethics to global challenges. A well-integrated interrelationship between liberal arts education on the one side, and specialized as well as professional education based in research on the other, is still the best proven method to propel Chinese universities into greater global prominence.

Some of these changes are already in evidence. For example, in 2004, Shanghai's Fudan University introduced a residential college structure, general education curriculum, and began to allow students to delay choosing a major until their second year. In 2009, Guangzhou's Sun Yat-sen University started experimenting with a liberal arts college for top students to study the Chinese classics, Greek and Latin, and the social sciences. Zhejiang University, Peking University's Yuanpei College, Tsinghua University, Southern University of Science and Technology, and Xingwei

College, among others, have introduced similar elements. This will only continue as regions that need to end dependence on manufacturing and build their future role in a knowledge-based economy invest in educating students who can best perform in that economy.

As they enter a period of change, Chinese universities have an opportunity, too, to invest in creating universities better designed to meet the needs of a global age. American institutions have over the course of a century assembled study abroad programs, exchanges, branch campuses, and other systems to extend their reach and influence internationally and to transmit to students "worldly knowledge"—that is, knowledge shaped by a broad set of cultural and national histories and conditions, with the potential to create more productive dialogue and collaboration on the kinds of fundamental global issues that will become increasingly critical in the years ahead.

Each of these systems has been useful, but each has limitations as well. Most recently, Berkeley has begun to develop a mutualist vision of the globalized university rooted in an assessment of the inexorable direction of the global future, which is increasingly knitted together around not just a single global research enterprise, but also the changing social and economic role of a preeminent research university. In contrast to the "high modernist" vision of the state university as a machine whose output would be knowledge workers contributing to the state economy, our recently announced Berkeley Global Campus—an internationally focused research and teaching hub being developed several miles from our main campus in partnership with other top global universities and private partners—represents the first-class research university as a focal point for enabling the state and its citizens to engage the world, connecting Berkeley scholars and local industry with researchers and innovators worldwide, and drawing human and financial capital from across the globe into the state. Rather than the cloistered space envisioned by the traditional inward-looking campuses, the Berkeley Global Campus will be a site for the flow of ideas, information, money, technology, and people moving not only between Berkeley and foreign universities, but also between the private and public sectors.

By acknowledging the irreversible force of global trends, the extent to which no local challenge is disconnected from global issues, and the powerful role that universities can play, we seek to establish a new kind of global presence that is in concert with our public mission. Any new conception of higher learning in China should, I believe, have similar elements embedded within it from the start.

With all this in mind, Western conceptions of liberal learning will no doubt be in some tension with China's current political system. The idea of connecting students to the world, exposing them to a broad education, and developing their skills in critical inquiry, while important for the economy, will also result in a more imaginative, creative, and doubtless questioning student body and citizenry. This will most likely produce conflicts with constraints on political freedom and academic freedom, while enhancing the quality of political debate in the larger context of an increasingly interactive, and interdependent, global marketplace.

A successful undergraduate education today is based in the foundational ideals and values of liberal learning as it has been articulated over the decades, while also evolving in adaptive ways to the demands of our time. Berkeley is seeking to ensure that it is a knowledge "community" rather than a "factory," that an undergraduate degree at Berkeley combines the best of what is available in liberal arts colleges with the resources of a great research university, offering courses that teach basic competencies while offering an almost unimaginable range of opportunities for specialization, exposure to research and professional fields, as well as chances to work with some of the best faculty in the world. Echoing Yale President Day in his Yale Reports, Berkeley takes on the obligation to cultivate intellectual curiosity, that is, not just to train our basic intellectual capacities to evaluate different ways of understanding and interpreting the world, but also to stimulate students to search relentlessly for new ways and approaches to acquire and advance knowledge. This does not mean teaching students a fixed curriculum, certainly in the manner advanced by Robert Maynard Hutchins at the University of Chicago in the mid-twentieth century,[8] but it does mean assuming the need for some fixed critical capacities. The faculty teach undergraduates not just so that they learn, but also so that they learn how to learn, whether on their own or in formal study. Increasingly, this means learning data numeracy as well as cultural literacy, worldly understanding as well as civic values, new skills for a rapidly changing world along with traditional values, habits, and dispositions.

As Berkeley builds a steadily proliferating architecture of academic offerings in our majors and specialized programs, it is working to maintain (and in some instances restore) a sense of common purpose in our undergraduate curriculum, as well as the importance of the extracurricular dimensions of college life (re-instating, for example, the importance of the residential college experience). Berkeley also seeks to balance the need to attain general knowledge with the need for students to have sufficient training for their lives after graduation, either in graduate or professional study or in high-level careers. Berkeley seeks as well to balance courses and training in the foundational principles of discrete disciplines with a range of applications that have robust practical implications. It is not an easy task; faculty must build curricular paths, moving students from general to advanced and specialized knowledge, in ways that can accommodate both wildly uneven levels of high school or community college preparation and the increasing technical, scientific, and intellectual challenges of almost every field. Here the twin credo of access and excellence is built into the undergraduate mission of the university in fundamental ways, since the diversity of the student body demands excellence in our undergraduate programs, especially when students need additional attention and, in some instances, remediation for inadequate high school training.

Berkeley also intends to ensure that all of our undergraduates learn to appreciate, and engage in, research. Research is not only an activity that should be reserved for graduate students, postdocs, and faculty, but can be made available as a resource for undergraduates at least in their latter two years. Research imparts skills that are specific to specialized projects while also teaching how to pursue knowledge on one's own. Research teaches scientific methodologies, and provides guided experiences in the use of libraries, special collections, archives, internet resources, community-based engagements, laboratory research projects, performance art, among many other pursuits. Research teaches how to measure the reliability, provenance, and character of sources: how to respect the importance of evidence, while knowing how evidence has been and can be used to different ends, and sometimes with multiple purposes.

Berkeley can settle for no less than to ensure that our undergraduates remain the full beneficiaries of the best set of undergraduate experiences available anywhere, in the larger context of the leading public research university in the world. To be sure of a future even brighter than our past, however, will require an education adapted to the needs of the new century and enlivened by global participation and scope. This will mean an increased reliance on institutions of higher education not only within the immediate locale (or that are part of the UC system of universities) but also in parts of the world such as China. As China invests in its own institutions and continues to expand its commitment to teaching and research, we can only benefit, even as we aspire to support and help advance China's capacities to join with us in confronting the world's most pressing challenges—all of which are global, and none of which we can solve on our own.

NOTES

1. A letter from Ezra Stiles to Eliphalet Williams, Dec. 3, 1777.
2. For more on the formation of land grant universities and their place in the modern American university system, Clark Kerr, *The Uses of the University*, (Harvard University Press, 1963), Chapter 1.
3. For background on this monumental piece of legislation and its impact, see Glenn C. Altschuler and Stuart M. Blumin *The GI Bill: A New Deal for Veterans* (Oxford University Press, 2009).
4. For an outline of the conditions leading up to the Master Plan and its early effects, see John Aubrey Douglass, *The California Idea and American Higher Education: 1850 to the 1960 Master Plan* (Stanford University Press, 2000).
5. See Nicholas Lemann, *The Big Test* (Farrar, Straus & Giroux, 1999), for a critical review of the idea of meritocracy in the context of the history of the University of California.
6. See Boyer Commission on Educating Undergraduates in the Research University, *Reinventing undergraduate education: A blueprint for America's research universities* (Carnegie Foundation, 1998).
7. For more on these sometimes competing and sometimes complementary ideas, see Nicholas Dirks, *Autobiography of an Archive: A Scholar's Passage to India* (Columbia University Press, 2015), p. 332; as well as Hanna Holborn Gray's 2009 Clark Kerr Lectures at the University of California, Berkeley.
8. For a study of Hutchins, his approach to general education, and his influence on the University of Chicago, see Mary Ann Dzuback, *Robert M. Hutchins: Portrait of an Educator* (University of Chicago Press, 1991).

REFERENCES

Delbanco, A. (2014). *College: What it was, is, and should be.* Princeton: Princeton University Press.

Dirks, N. B. (2015). *Autobiography of an archive: A scholar's passage to India.* New York: Columbia University Press.

Geiger, R. L. (2014). *The history of American higher education: Learning and culture from the founding to World War II.* Princeton: Princeton University Press.

Harvard University. (1885). *Report of the president of Harvard College and reports of departments.* Published by Harvard University, Cambridge, Mass.

Kerr, C. (1963). *The uses of the university.* Cambridge, MA: Harvard University Press.

Missouri State Board of Agriculture. (1866). *Annual report of the Missouri State Board of Agriculture.* https://babel.hathitrust.org/cgi/pt?id=hvd. hnyplb;view=1up;seq=5

Packard, A. S. (1827). The substance of two reports of the faculty of Amherst College to the board of trustees, with the doings of the board thereon. *North American Review, 28,* 300.

President's Commission on Higher Education. (1947). *Higher education for American democracy.* Washington, DC: US Government Printing Office.

Robson, D. W. (1985). *Educating republicans: The college in the era of the American Revolution, 1750–1800* (Vol. 15). Westport: Greenwood Publishing Group.

Röhrs, H. (1987). The classical idea of the university. In *Tradition and reform of the university under an international perspective.* New York: Peter Lang International Academic Publishers.

Tappan, H. P. (1851). *University education.* New York: GP Putnam.

Tucker, T. L. (1979). Centers of sedition: Colonial colleges and the American Revolution. *Proceedings of the Massachusetts Historical Society,* Third Series, *91,* 16–34.

Yale College. (1828). *Report of the course of instruction in Yale College by a committee of the corporation and the academical faculty.* New Haven: Yale College.

CHAPTER 9

Transcending Boundaries: Educational Trajectories, Subject Domains, and Skills Demands

Dirk Van Damme

Abstract This chapter reviews the links between new skills demands, higher education curriculum developments, and the growing interest in liberal arts and science education in various parts of the world. It assesses the value of a liberal arts and sciences model in respect of the changing skills demand from the labour market, the need to (re-)connect various knowledge domains that transcend the traditional disciplines and thus consider curriculum reform, and the link with trends and reforms in secondary education. Liberal arts and sciences programmes can be seen as a possibly powerful response to global demands and challenges which many countries are facing.

Keywords higher education curriculum • skills • learning outcomes • disciplines

D. Van Damme (✉)
Organization for Economic Co-operation and Development, Paris, France

Ghent University, Gent, Belgium

INTRODUCTION

The global development of liberal arts and sciences programmes in higher education raises important and fascinating questions with regard to the linkages of higher education curricula and pedagogy to the world of work and wider social benefits. Are liberal arts and sciences programmes providing a positive answer to the needs in the knowledge economy for creativity, critical thinking, and innovation skills? Do they offer alternative outlooks on the organisation of human knowledge, more suited to meet the demands of contemporary scientific research and technological development? Can they be seen as a move 'back to the future' of the very early roots of the university, in an economic and social context where many other boundaries are vanishing?

These questions go well beyond what can possibly be answered in this chapter. But they inspire the reflection on the links between new skills demands, higher education curriculum developments, and the growing interest in liberal arts and sciences education in various parts of the world. Despite their historical roots, liberal arts and sciences programmes still seem to be a field of experimentation and systemic experiential learning on how the higher education system can respond better to the social demands for innovation and excellence. Notwithstanding its diffuse and diverse expansion and its lack of homogeneity, the liberal arts and sciences education movement across the globe should be seen as a major force of innovation and reform in higher education. It is certainly not the only one and there are many equally interesting reform initiatives in higher education, but the liberal arts and sciences education is the one that forces us to think deeply about higher education's future.

WHAT DO STUDENTS LEARN?

One of the most intriguing questions about higher education is what exactly causes students to transform during the time they spend in colleges and universities in such a way that they take such a huge benefit from that experience in later life?

The evidence on the positive outcomes of higher education is overwhelming and very stable over time. The employment rate of 25- to 64-year-olds with a higher education degree is some 10 percentage points higher than for individuals with only an upper secondary education. The earnings premium for higher educated individuals compared to colleagues with only upper secondary education is estimated to be around 60%. Over a lifetime, the net present value of higher education is around US$160,000 for a man and US$100,000 for a woman on average across OECD countries. Higher education also results in important social benefits in health, social capital, interpersonal trust, or political participation. For example, 88% of higher educated individuals (self-)report good health against 79% of people with an upper secondary education (OECD 2014). These benefits are substantial and, as far as we know, have not been eroded by the economic crisis, quite on the contrary. They also seem to last over one's lifetime and do not diminish with age.

But we know very little about the processes that produce these outcomes. We do not understand well how the student's experience in university, the quality of the teaching and learning environment, the student's learning effort, and other variables interact in producing the added value in knowledge, skills, character, and metacognition that transform an individual into an educated person whom employers are willing to employ and to whom societies distribute many other scarce goods.

The traditional human capital theory centres on the substantive contribution of the teaching and learning process in higher education institutions to knowledge, skills, and other attributes of students. Unfortunately, we have very few data on the skills equivalent of higher education qualifications. The OECD Survey of Adult Skills (PIAAC) is the first attempt to measure skills among adult populations and its data allow the comparison of literacy and numeracy skills levels among higher education graduates (OECD 2013b). The data on skills of 20- to 34-year-old higher education graduates show two important points. First, the variation between countries is huge. The mean literacy score ranges from 291 (Italy) to 333 (Finland). The mean of the top-performing countries is higher than the 75th percentile of some of the low performers. Some of the countries with many universities featured in the top of the academic rankings demonstrate only very mediocre skills outcomes, while some countries more peripheral in the global academic system perform much better. Second, also within countries there is a lot of variation in the literacy score of young people with a university degree. A higher education degree does not seem to narrow the skills distribution very much.

Sure, literacy skills are foundational skills which are not primarily supposed to be acquired in higher education. Teaching and learning in universities have probably a higher value-added in more specialised skills sets. But these data are worrisome to the extent that they dismiss the idea that higher education qualifications signal a certain threshold skill level or guarantee employers a minimum skill set. In fact, the overlap in the distribution of literacy skills between university graduates and individuals with only an upper secondary education as the highest level of education is very large. This means that the discriminatory function of a university degree is very much flawed. Hence, it is almost very problematic to regress the economic and social benefits of a higher education degree to its added value in terms of skills development.

The traditional human capital view is increasingly challenged by the 'signalling' or 'screening' hypothesis, which emphasises the selective functions of university programmes in providing employers with workers fit for specific job categories. Higher education thus acts as a mechanism saving employers the hassle of expensive recruitment, selection, and testing to identify the workers they need. In this hypothesis, the added value of a higher education plays a negligible role in the allocation of graduates to jobs, earnings, and other status goods.

For the screening or signalling hypothesis to work well, universities need to have some form of monopoly in awarding degrees which open doors to jobs. In most countries, this still seems to be the case. Recent OECD analysis (Paccagnella 2015) demonstrated that across countries participating in the OECD Survey of Adult skills (PIAAC) earnings are much more driven by formal education than by actual levels of skills. Not only the institutional regulation of labour markets but also the high symbolic power of university degrees ensure that qualifications, not skills, determine access to high-level jobs and earnings.

The contradiction between the high symbolic value of a higher education degree and their low value in terms of signalling a certain threshold level of skills might become one of the main systemic challenges for higher education. Recent research has provoked doubts on what students exactly learn in college and the added value in terms of knowledge and skills of students' higher education experience. In his controversial book *Our underachieving colleges* former Harvard president Derek Bok (2008) critiqued American colleges for not providing students with enough added value in developing writing, critical thinking, quantitative and moral reasoning skills. Arum and Roksa (2011) analysed data from the Collegiate Learning Assessment (CLA) instrument administered to a large sample of undergraduate students and concluded that a large share of them demonstrated no significant improvement in a range of academic skills during the first two years of college. In a subsequent report (Arum et al. s.d.), they confirmed and expanded their findings to conclude that 'large numbers of college students report that they experience only limited academic demands and invest only limited effort in their academic endeavours'.

Comparative international evidence on the learning outcomes of higher education graduates still is very fragmentary. OECD's Assessing Higher Education Learning Outcomes (AHELO) project so far has not moved beyond the feasibility study (OECD 2013a) and its large scale implementation meets strong resistance from parts of the higher education community. Still, in many countries the issue of higher education students' learning outcomes is high on the agenda and several interesting national and international research initiatives have been initiated to gather evidence and further insights into the matter (Van Damme 2015).

There are clear signals that employers are also becoming less and less satisfied with the kind of selection and training services higher education institutions are providing. They seem to be increasingly unhappy about the level and substance of graduates' skills and they start to rely more on their own human resources management capacity for selecting and testing candidates. In some cases they exhibit manifest distrust to universities' ability to test and select the best students. An interesting recent example was the announcement of Ernst & Young in the UK that it would no longer look at higher education qualifications in the selection of candidates (Havergal 2015).

Employers' organisations complain about the lack of twenty-first-century skills such as problem-solving, team work, or communication and the limited employability of graduates. Of course, this is not new, and probably also inevitable. Universities define their mission broader than employability. Fast changes in technology and innovation of work processes and management structures imply that skills of even recently graduated workers rapidly become obsolete. Employers and professional organisations tend to shift the blame for skills gaps to universities, to save themselves from expensive professional development and reskilling programmes.

But, having said this, the transformation of the global economy after the Great Recession seems to accelerate changes in skill demand, especially with the innovative firms at the top of the global value chains. Universities seem to be too slow in adjusting their educational programmes, and a widening skills gap, especially at the top of the distribution, is the consequence. For example, recent work at the UK's National Centre for Universities and Business suggests that global companies feel badly served by universities in providing them with top-level students with the right kind of skills for an innovative, global environment they will be working in (Stevenson 2014). Half of the companies surveyed indicated that they already introduced their own innovative strategies to find and develop talent.

SKILLS DEMANDS OF INNOVATIVE ECONOMIES
AND SOCIETIES

Employers' dissatisfaction raises the question about the kind of skills that higher education would need to foster to meet the demands of innovative economies and societies. Changing skills demand has to be understood against the transformation of employment and work as part of the global economic development. Technological change is an important driver of changing skills demand, with new technologies tending to be biased towards higher skills (Goldin and Katz 2008). Further research has shown a pattern of more polarised shifts in employment leading towards a U-shaped labour market, where also the low end of the skills distribution gets a higher share of labour market and occupations in the middle low employment shares. The so-called skill-biased technological change drives economies towards ever higher skills needs and, hence, might explain why higher numbers of higher education graduates are easily absorbed into labour markets and why benefits and returns—at least comparative to middle-educated workers—remain high.

Still, even if machines have not replaced human labour in the past, it is difficult to predict the future. Polarisation in the labour market might well not continue very far in the future (Autor 2015). With the frontiers of automation advancing at a rapid pace, the employment future of higher education graduates might become uncertain, certainly for those who are predominantly trained for routine cognitive tasks characterised by repetition and predictability, such as bookkeeping, data entry, and similar procedural tasks. Routine, codifiable tasks will be easily digitised and automated, even those tasks now performed by highly trained professionals. Instead, non-routine tasks mobilising workers' problem-solving, flexibility, communication, and creativity skills will become more important (Autor et al. 2003; Autor and Price 2013).

So, the issue is not so much the level of future skills demand, but the nature of skills which graduates will have to master to be employable and contribute to innovation, productivity, and growth. OECD's work on the skills needed for innovation has tried to elucidate that question. On the basis of an analysis of international data sources it was shown that the most innovative professionals had a varied mix of academic qualifications, including—not surprisingly—engineering and sciences, but also business, social science, and even arts degrees (Toner 2011; Avvisati et al. 2013). The traditional view that innovators mainly come from science and engineering is partly confirmed, but the analysis also suggests that innovation benefits from a much larger set of specialisations and that the boundaries between academic fields of study are less important in the world of work than they seem to be in academia.

Furthermore, the analysis of Avvisati et al. (2013) also sheds light on the skills that were expected from innovative professionals in innovative jobs and on the skills that distinguished them from non-innovators. Highly innovative professionals have higher job requirements for any single skill than non-innovative workers, but their jobs also appealed much more to creativity, presentation, the 'alertness to opportunities', analytical thinking, the 'ability to coordinate activities', and the 'ability to acquire new knowledge' than was the case for non-innovative workers. This analysis provides rather rare empirical support for the importance of so-called '21st century skills' for innovative jobs and industries. Skills such as creativity and critical thinking, problem-solving, making connections, complex communication skills, teamwork, flexibility, and global competences are now generally seen as critically important for the workplaces and labour markets of tomorrow.

Added to those are skills that can be labelled as 'character' or social and emotional skills. The Center for Curriculum Redesign (CCR) has recently tried to develop a comprehensive 'character qualities framework', in which these social and emotional skills are identified and integrated (CCR 2015). It has come up with the following list: mindfulness, curiosity, courage, resilience, ethics, and leadership. Also these behavioural attributes play an important part in equipping young people for the jobs and social responsibilities in the world of tomorrow. They strongly overlap with the skills identified by global companies as necessary for future leaders to confront ever more complex challenges in an uncertain and volatile environment.

There are several attempts to integrate the various dimensions of skills demands and expectations of future graduates into one comprehensive framework, useful to guide curriculum reform. CCR quite convincingly argues for a four-dimensional framework of knowledge, skills, character, and metacognition: 'Knowledge must strike a better balance between traditional and modern subjects, as well as interdisciplinarity. Skills relate to the use of knowledge, and engage in a feedback loop with knowledge. Character qualities describe how one engages with, and behaves in, the world. Metacognition fosters the process of self-reflection and learning how to learn, as well as the building of the other three dimensions' (CCR 2015).

SUBJECT FIELDS: OLD BOUNDARIES AND NEW INTERCONNECTIONS

It is quite interesting to note that the mastery of a specific disciplinary field of study is not identified as one of the very top skills that differentiate innovative from non-innovative professionals (Avvisati et al. 2013). This seems to suggest that the specific field of study is not very important for the contribution a skilled worker can make to innovation. And, at a superficial look, this might also contradict the current policy concerns for the low numbers of Science, Technology, Engineering, and Mathematics (STEM) graduates that are seen as critically important for innovation, productivity, and growth.

On an average, across OECD countries, 15% of new higher education entrants choose 'engineering, manufacturing and construction', and another 20% 'science, life sciences, math and computing', against 20% for 'humanities and education', and 31% for 'social sciences, business and law' (OECD 2014). But in many countries, the numbers of STEM entrants are much lower than the average OECD figure. The number of STEM degrees has increased over the past years, but the rate of increase was lower than the increase in higher education degrees in general, so the relative share of STEM graduates in the population with a higher education has gradually declined.

Another interesting observation made in recent OECD research is that the 'field-of-study' mismatch between graduation and employment is actually rather high, also for STEM graduates (Montt 2015). Across countries that participated in the OECD Survey of Adult Skills no less than 65% of workers trained in 'science, life sciences, math and computing' are actually working in another field than the one they have been trained for, much higher than the average mismatch across fields of study of 39%. The figure is not much lower than the one for 'humanities and education' of 73%, but much higher than the one for 'social sciences, business and law' of 23%. For 'engineering, manufacturing and construction' it is 33%. These data nuance the widespread concern about low numbers of STEM graduates as being not only a problem of the choice of study at the entry of higher education, but also of the suitable employment opportunities afterwards. But they also indicate the importance of the post-gradation professional mobility on the labour market.

A recent report on the STEM workforce of the US National Science Foundation (NSF 2015) has further qualified the debate. It requests policymakers to move away from a narrow focus on STEM qualifications and to better understand the heterogeneity of the STEM workforce. It also demonstrates that STEM workers do not follow one linear trajectory, but that there are multiple pathways leading to STEM jobs. At least in the USA—but if we look at Montt's data, the same is probably true for other countries—there are rather loose links between field of study and actual STEM occupations. STEM workers can be trained in disciplines which are not directly identified as STEM. The report argues for a more holistic approach to STEM education, an approach by which across many disciplines, a STEM-'capable' workforce is trained. Ultimately, all future workers need to have access to high-quality higher education that enables them for STEM capability. Liberal arts and sciences programmes have understood this very well.

This is very much in line with Avvisati et al. (2013) when they say that many of the critically important skills for innovation can be fostered in all fields of study, even if it could take a different shape from one subject to another. STEM graduates might be in high demand, not because of their specific, technical skills, but because of their strengths in innovation skills. Employers reward the critical thinking, problem-solving, behavioural and social skills perceived to be part of STEM education, rather than the technical STEM skills.

Whether similar arguments can be made for other fields of study is less clear. For example, for the arts—a field of study of which a significant number of graduates end up in innovative jobs five years after graduation, especially in product innovation (Avvisati et al. 2013)—a review of experimental research in arts education in schools demonstrated no real significant effect on various other cognitive and non-cognitive skills, other than through selection effects (OECD 2013c). The specific contribution of various disciplines and fields of study to the development of innovation skills or 'twenty-first-century skills' is still largely unchartered territory.

Anyway, the discussion so far seems to provide support for smart new combinations of disciplines in higher education curricula. Interdisciplinarity might have become a fashionable and often superficial mode of curriculum reform (Jacobs 2014), but interdisciplinarity also is a core component of many interesting examples of curriculum and pedagogical reforms in higher education, such as problem-based learning (Hoidn and Kärkkäinen 2014). Disciplines go back to the eighteenth- and nineteenth-century organisation of human knowledge and have institutionalised themselves in various ways. But today the most fascinating discoveries and frontier developments in scientific research are to be found at the boundaries or in the intersections of disciplines. Interdisciplinarity should not be understood as simply mixing multiple disciplines, but as a smart way to spell out the interconnectivity between various complementary viewpoints and a necessary condition to solve today's complex problems.

LIBERAL ARTS AND SCIENCES EDUCATION AS CURRICULUM REFORM

The obvious question after the discussion so far then is how these developments impact the curriculum and its reform in higher education. The field of curriculum studies is predominantly concentrated on school education and is not yet well developed in higher education (Vidovich et al. 2012). And when curriculum reform in higher education is studied, most of the attention goes to the formal processes and characteristics of reform initiatives, and rarely to the substance and direction of reform.

At the same time, a lot is going on in higher education and many institutions are reshaping and redesigning their teaching and learning environments to optimise the student learning experience (Kärkkäinen 2012; Pegg 2013). After a period in which the global higher education community was focusing on improving research performance and getting integrated in the global research system, universities are now turning their institutional energy and resources to improving teaching and learning. Governmental initiatives have been taken to support the improvement and innovation of teaching and learning, while also regional and global reform processes, such as the Bologna Process and its companion Tuning Project, are inducing change and reform in the design of higher education programmes. This includes for example, reviewing the programme's learning objectives, rethinking pedagogies, choosing the right learning resources, or changing student assessment. Technology also is a great driver of reform in curricula and pedagogy (Sharples et al. 2014).

Curricula and their translation into content and pedagogical design can be seen as a particular and time/space specific organisation of knowledge, skills, character, and metacognition with the following characteristics: (1) a curriculum balances and integrates external demand (demand side) and internal purpose and mission (supply side); (2) it has to be economic in terms of resources and effective use of students' learning time, so it includes a normative framework that helps to distinguish and select the important from the trivial; (3) it integrates a psychological framework of learning, ideally based on the current state of learning sciences; and (4) it includes a theory of action, which describes how content and pedagogy help to realise the intended learning outcomes of students that meet the external demands and internal purposes. Obviously, in most cases a lot of all this remains implicit; institutions are very rarely open and transparent about what drives them in curriculum reform. But, essentially, all curriculum reform efforts are intentional in the way they try to shape the students' teaching and learning experience and the learning objectives that should result from it.

Curricula are set in a specific time and space configuration, but at the same time also transcend time and space. Contemporary higher education curricula still are strongly located in local and national spaces, but probably have stronger global ambitions than in the past. And they carry with them a heavy historical legacy, but increasingly aim for the future in their ambitions to prepare students for a lifelong career and a fulfilling personal life, and in their ambitions to serve humanity's future and mankind's resilience against many challenges.

Curriculum reform to a large extent is a process of selection. First of all, a selection has to be made of the most relevant external demands with which a programme is confronted. In answering the question 'what should students learn' a myriad of external demands rival for attention and importance. Second, a selection has to be made in the purposes of the programme and the learning objectives derived from them. Third, a selection then is necessary of the best-suited content and pedagogical approaches to realise these learning objectives. The need to be selective implies that often curricula reform approaches are value-loaded, because the arguments with which these choices are justified are normative in kind.

These few short conceptual notes are necessary to understand that any curriculum reform approach is a specific attempt to answer these questions and meet these requirements. For example, the 'Problem Based Learning' approach is a particular way to organise the curriculum, content, and pedagogy, drawing on insights on student-centred learning, and on reflections on what really matters for students' learning outcomes and professional competence (Kärkkäinen 2012). In much the same way, we can see the liberal arts and sciences education as a specific response to the curriculum reform pressures in higher education. The implementation in specific contexts and institutions might differ, but there are strong unifying elements that can be described along the following lines:

Well-Roundedness A characteristic feature of liberal arts and sciences education is its case for well-roundedness. This concept refers to very old academic traditions, but has also references to very different cultural contexts. It basically denotes the importance of mastery of very different areas of knowledge and understanding, but also of the harmonious balance between knowledge and other forms of human understanding including the arts and ethics. The concept reminds of Howard Gardner's (2011) views on truth, beauty, and goodness as virtues in education. The harmonious development of the person living in the twenty-first century is an important objective of liberal arts and sciences programmes.

Connecting Tradition, Modernity, and Innovation Liberal arts and sciences education has strong roots in the very old traditions of the European university, although more utilitarian approaches and the Humboldtian model of the research university contributed to its sharp decline in Europe (Van der Wende 2011). Interestingly, the model now also appeals to very old traditions in various Asian educational contexts, as several chapters in this book explain. At the same time the continued existence (in the USA), its resurgence (in Europe), and its growing popularity (in Asia) show how well the model integrates with modernity and the modernisation imperative in developed and emerging countries alike. Yet, its pedagogical objectives and ambitions also align very well with the arguments in favour of twenty-first-century skills, as can be seen in attempts to focus on critical thinking, creativity, and problem-solving.

Intercultural Especially new developments in liberal arts and sciences education in Europe (Van der Wende 2011) have a very clear focus on interculturality and a global approach, both in their international student intake, the use of English as the language of instruction, and the global embedding of curricula content. Global competences such as a profound understanding and respect for cultural diversity and interdependence are an integral part of these programmes' learning objectives.

Interdisciplinarity A common, defining feature of liberal arts and sciences programmes is their integration of various disciplines into a coherent curriculum, often combining natural and life sciences, social sciences, and humanities and the arts into interesting and innovative combinations, with large numbers of elective courses open to students to construct their own curriculum. At least in Europe, but probably also elsewhere, the (re-) emergence of liberal arts and sciences education coincides with a more general tendency to avoid early specialisation and to build a more comprehensive and integrated curriculum at the undergraduate level.

Diversification for Excellence Several newly established liberal arts and sciences programmes take a more selective approach than usual in their systems, as is the case for example in the Netherlands. It is a clear ambition to be a component of diversification in the system, whereby specific and intellectually challenging trajectories are offered to very talented students. Opening routes to excellence is seen as a corrective against the confusion prevalent in many systems that massification equals mediocrity. Trajectories towards excellence also function as pathways for future leaders, trained in a global environment and educated through highly demanding tasks.

Matching Educational Trajectories Liberal arts and sciences education not only has the ambition to form well-rounded persons with a breadth of experience and knowledge, but also aims to be a key component of a harmonious educational trajectory over the course of life. Liberal arts and sciences programmes are thinking very critically—much more than usually in higher education—about how they can position themselves between secondary school education, postgraduate studies, entry into professional careers, and lifelong learning. They might well be at odds with some tendencies that can be observed in school education, such as the decline in arts education or increasing specialisation, but they are also in line with attempts to modernise the school curriculum, for example in innovative schools and learning environments (OECD 2013d) or initiatives like the International Baccalaureate. Through their curricula they also ensure that students have broad options for postgraduate study. Several liberal arts and sciences programmes include provisions for educational career guidance and invite students to think critically about their educational choices and life trajectory. In contrast to Zakaria (2015) in his powerful defence of liberal education, I would not oppose liberal arts and sciences education against more utilitarian and vocational purposes and objectives in higher education. The interesting thing is that it aims at combining and integrating vocational and general academic objectives.

Conclusion

The growing interest for liberal arts and sciences education across countries and cultures today, demonstrated by this book, suggests that it should not so much be seen as the product of specific historical and cultural circumstances and forces, but as a possibly powerful response to global demands and challenges which many countries are facing. The specific form liberal arts and sciences programmes take in particular settings might well be moulded by the local and national history and culture, but the common thread across the globe is probably that such programmes hold the promise to provide better answers to needs and demands of the twenty-first century than some of the competing curriculum reform models.

One of the reasons for this is that liberal arts and sciences education in many ways aims at 'transcending boundaries', an ambition that is very appealing in the modern world. Future leaders will have to be able to overcome the many divisions, boundaries, and segmentations of today's world in order to solve its problems. Bridging disciplines, spheres of life, old and new skills demands, cultural and political cleavages, and so on is very much at the heart of liberal arts and sciences education. If we want to understand why liberal arts and sciences education seems to appeal to many people, higher education institutions, and students in very different parts of the world, this may be one of the answers.

References

Arum, R., & Roksa, J. (2011). *Academically adrift: Limited learning on college campuses.* Chicago: Chicago University Press.

Arum, R., Roksa, J., & Cho, E. (s.d.). Improving undergraduate learning: Findings and policy recommendations from the SSRC-CLA longitudinal project (SSRC). Retrieved August 13, 2015, from https://s3.amazonaws.com/ssrc-cdn1/crmuploads/new_publication_3/%7BD06178BE-3823-E011-ADEF-001CC477EC84%7D.pdf

Autor, D. H. (2015). Why are there still so many jobs? The history and future of workplace automation. *Journal of Economic Perspectives, 29*(3), 3–30.

Autor, D. H., & Price, B. (2013). The changing task composition of the US labor market: An update of Autor, Levy, and Murnane (2003). Unpublished Manuscript. Retrieved October 28, 2015, from http://economics.mit.edu/files/9758

Autor, D. H., Levy, F., & Murnane, R. J. (2003). The skill content of recent technological change: An empirical exploration. *The Quarterly Journal of Economics, 118*(4), 1279–1333.

Avvisati, F., Jacotin, G., & Vincent-Lancrin, S. (2013). Educating higher education students for innovative economies: What international data tell us. *Tuning Journal for Higher Education, 1,* 223–240.

Bok, D. (2008). *Our underachieving colleges. A candid look at how much students learn and why they should be learning more.* Princeton: Princeton University Press.

Center for Curriculum Redesign. (2015). *Character education for the 21st century: What should students learn?* Boston: CCR. Retrieved August 25, 2015, from http://curriculumredesign.org/wp-content/uploads/CCR-CharacterEducation_FINAL_27Feb2015.pdf

Gardner, H. (2011). *Truth, beauty and goodness reframed. Educating for the virtues in the twenty-first century.* New York: Basic Books.

Goldin, C., & Katz, L. (2008). *The race between education and technology.* Cambridge, MA: Harvard University Press.

Havergal, Chr. (2015), Ernst and Young drops degree classification threshold for graduate recruitment. *Times Higher Education.* Retrieved August 25, 2015, from https://www.timeshighereducation.com/news/ernst-and-young-drops-degree-classification-threshold-graduate-recruitment

Hoidn, S., & Kärkkäinen, K. (2014). *Promoting skills for innovation in higher education: A literature review on the effectiveness of problem-based learning and of teaching behaviours* (OECD education working papers, no. 100). Paris: OECD Publishing. doi:10.1787/5k3tsj67l226-en.

Jacobs, J. A. (2014). *In defense of disciplines. Interdisciplinarity and specialization in the research university.* Chicago: Chicago University Press.

Kärkkäinen, K. (2012). *Bringing about curriculum innovations: Implicit approaches in the OECD area* (Vol. OECD education working papers, no. 82). Paris: OECD Publishing. doi:10.1787/5k95qw8xzl8s-en.

Montt, G. (2015). *The causes and consequences of field-of-study mismatch: An analysis using PIAAC* (OECD social, employment and migration working papers, no. 167). Paris: OECD Publishing. doi:10.1787/5jrxm4dhv9r2-en.

National Science Foundation. (2015). *Revisiting the STEM workforce. A companion to science and engineering indicators 2014.* Arlington: NSF. Retrieved August 26, 2015, from http://www.nsf.gov/nsb/publications/2015/nsb201510.pdf

OECD. (2013a). *Assessment of Higher Education Learning Outcomes (AHELO): Feasibility study report. Volume 2—data analysis and national experiences.* Paris: OECD Publishing.

OECD. (2013b). *OECD skills outlook. First results from the survey of adult skills.* Paris: OECD Publishing. http://skills.oecd.org/OECD_Skills_Outlook_2013.pdf.

OECD. (2013c). *Art for art's sake? The impact of arts education.* Paris: OECD Publishing.

OECD. (2013d). *Innovative learning environments.* Paris: OECD Publishing.

OECD. (2014). *Education at a glance 2014. OECD indicators.* Paris: OECD Publishing. http://www.oecd.org/edu/Education-at-a-Glance-2014.pdf.

Paccagnella, M. (2015). *Skills and wage inequality: Evidence from PIAAC* (OECD education working papers, no. 114). Paris: OECD Publishing. doi:10.1787/5js4xfgl4ks0-en.

Pegg, A. (2013). *"We think that's the future": Curriculum reform initiatives in higher education.* Heslington: The Higher Education Academy. Retrieved August 25, 2015, from https://www.heacademy.ac.uk/sites/default/files/curriculum_reform_final_19th_dec_1.pdf

Sharples, M., Adams, A., Ferguson, R., Gaved, M., McAndrew, P., Rienties, B., Weller, M., & Whitelock, D. (2014). *Innovating pedagogy 2014: Open university innovation report 3.* Milton Keynes: The Open University.

Stevenson, M. (2014). *Developing exceptional talent. The education of global leaders.* NCUB green paper nr. 2. London: NCUB. Retrieved August 25, 2015, from http://www.ncub.co.uk/index.php?option=com_docman&task=doc_download&gid=147&Itemid=

Toner, P. (2011). *Workforce skills and innovation: An overview of major themes in the literature* (OECD education working papers, no. 55). Paris: OECD Publishing. doi:10.1787/5kgk6hpnhxzq-en.

Van Damme, D. (2015). Global higher education in need of more and better learning metrics. Why OECD's AHELO project might help to fill the gap. *European Journal of Higher Education, 5*(4), 425–436. doi:10.1080/2156823 5.2015.1087870.

Van der Wende, M. (2011). The emergence of liberal arts and sciences education in Europe: A comparative perspective. *Higher Education Policy, 24,* 233–253.

Vidovich, L., O'Donoghue, T., & Tight, M. (2012). Transforming university curriculum policies in a global knowledge era: Mapping a "global case study" research agenda. *Educational Studies, 38*(3), 283–295. doi:10.1080/0305569 8.2011.598681.

Zakaria, F. (2015). *In defence of a liberal education.* Norton: WW Norton & Company.

INDEX

Notes: Page references with 'n' denotes notes

© The Editor(s) (if applicable) and The Author(s) 2016   143
W.C. Kirby, M.C. van der Wende (eds.), *Experiences in Liberal
Arts and Science Education from America, Europe, and Asia,*
DOI 10.1057/978-1-349-94892-5